Safety Management

T0330744

Recent work has demonstrated that incidents, accidents and disasters tend to result from complex socio-technical failures, rather than just 'human error' on the one hand, or simple technical failures on the other. For the reduction of accidents, therefore, it is necessary to deal with systems factors, in which both technical and human-factors elements play an equal and complementary role. However, many of the existing techniques in ergonomics and risk management concentrate on plant/technical issues and downplay systems factors and 'subjectivity'. The present text describes a body of theory and data which addresses this issue squarely, drawing on systems theory and applied psychology, and which stresses the importance of human agency within systems. The central roles of social consensus and reliability, and the nature of verbal reports and 'functional discourse' are explained in some detail.

This book therefore presents a new 'Qualitative Systems Approach' to safety management, offering both greater safety and economic savings. It presents a series of methodological 'tools' whose reliability and validity have been shown through extensive work in the rail and nuclear industries and which allow organisational and systems failures to be analysed much more effectively in terms of quantity, precision and usefulness.

This is a textbook for undergraduate and graduate students in occupational psychology, human factors, ergonomics and HCI, and the sociology of disasters and risk. It is also useful for safety managers and professionals in many safety critical firms and organisations, reliability engineers, risk managers, and human factors specialists.

Safety Management
A qualitative systems approach

John Davies, Alastair Ross, Brendan Wallace and Linda Wright

CRC Press
Taylor & Francis Group
Boca Raton London New York

CRC Press is an imprint of the
Taylor & Francis Group, an **informa** business

CRC Press
Taylor & Francis Group
6000 Broken Sound Parkway NW, Suite 300
Boca Raton, FL 33487-2742

First issued in paperback 2019

© 2003 by Taylor & Francis Group, LLC
CRC Press is an imprint of Taylor & Francis Group, an Informa business

No claim to original U.S. Government works

ISBN-13: 978-0-415-30370-5 (hbk)
ISBN-13: 978-0-367-39521-6 (pbk)

Visit the Taylor & Francis Web site at
http://www.taylorandfrancis.com

and the CRC Press Web site at
http://www.crcpress.com

Contents

Figures

Tables

Preface

The ideas in this book were first developed in the late 1990s during our work with the railway industry (CIRAS: Confidential Incident Reporting and Analysis System), and the nuclear industry (SECAS: Strathclyde Event Coding and Analysis System). In the railway industry, as we developed CIRAS and analysed the reports, we discovered to our surprise that we were gaining very little information from staff about what was classically referred to as 'human error'. Instead, they gave us data relating to systems features, organisational problems and sociological aspects of the industry, producing a picture very far from the classic 'individualistic' view of a single operator committing an 'error'. At the same time, in our work with the nuclear industry, we were finding that a key concept in error taxonomies is 'reliability'. However, it took us some time to realise that the way we were using the word (as social scientists) was very different from the use of the word in the engineering context. Consequently, taxonomies were not being properly tested for reliability (in the sense that we were using it: agreement between users on individual classifications), which seemed to be viewed as a trivial part of the validation process that could easily be dispensed with.

It wasn't until some time later that we saw that these issues were two sides of the same coin, and it was as a result of this that we were led to question the 'cognitivist' paradigm which has often been used as the main psychological model in safety management and human factors. We started to think that perhaps 'cognitivism' was holding back the increasing (and welcome) developments in the field that emphasised systems features of the accident and error process. One of the key problems with ergonomics, we started to think, was that writers who emphasised the systems aspects of 'error' also tended to emphasise 'cognitivist' aspects, despite the fact that, in our opinion, these two viewpoints were incompatible.

In search of an 'alternative paradigm' that matched the reality of what we were finding in our research work, we started on an exciting exploration of the new 'post-cognitivist' work that has been done in psychology in the last twenty years. What was particularly gratifying in this 'voyage of discovery' was discovering the links between apparently disparate thinkers. Studying 'systems' theory and cybernetics led us forward to connectionism and 'dynamicism' but it also led us back to hermeneutics and phenomenology. And this in turn led us to new views of epistemology, which led to a redefinition of the word reliability, and a restatement, within our new 'paradigm', of why it was so important. We discovered links that we had never dreamed existed: between Continental philosophy, ordinary language philosophy, and the most up-to-date, 'cutting edge' research of the Artificial Life, and Parallel Distributed Processing (PDP) communities.

At the end of this process we had 'woven' a distinctive approach to human error, accidents, disasters, error taxonomies and so on. But, at the same time we had discovered, not just new ideas (or new to us, at least!), but new methodologies. The reader will, therefore, hopefully discover thematic development in the book, in that it generally moves from the theoretical to the practical. The first few chapters explain the 'philosophy' that lies behind our approach, before we move onto the specific projects that led to our development of alternative methods and techniques. There then follow discussions of some 'traditional' philosophies in the field and explanations as to why they have not always functioned as well as they might have been expected to. Finally we conclude with new, practical solutions to traditional problems.

As the title of this book suggests, our approach is best described as a Qualitative Systems Approach (QSA). It is qualitative, because we feel that researchers should pay attention to the discourse and language used by the actual managers and staff trying to lower risk and increase safety. However, despite the fact that it is individuals that use language, we should never forget that these individuals exist in an organisation (or a system), and that their behaviour is constrained by and constructed within this system. It is only by continually moving ('dialectically') from the individual to the system and back again, from the discourse and behaviour of individuals to the rules and structures of the company or organisation, that a full view of safety practices (and shortcomings in these practices) can be given.

We should conclude by mentioning that the main absence in this text is reference to our debt to the Pragmatists: John Dewey, William James and, of course, C. S. Peirce. However, we feel that to make specific references would miss the point; we would hope that the pragmatic spirit animates the whole volume. We have attempted to create a volume of practical solutions to practical problems. We hope that after finishing the book, the reader will feel we have succeeded.

Notes on contributors

Professor **John Davies** is director of the Centre for Applied Social Psychology (CASP) at the University of Strathclyde in Glasgow. He has an extensive research and publication record in the areas of Human Factors and Public Health. He has contributed widely to national and international conferences on both these topics. He is also director of Human Factors Analysts Ltd. (HFAL).

Alastair Ross is a Senior Research Fellow in the Centre for Applied Social Psychology (CASP) at the University of Strathclyde. He has developed taxonomies and coding systems for a number of high-consequence industries. He has an interest in discourse and attribution, in particular the psychology of sport and the way in which natural explanations of élite athletes relate to subsequent performance.

Dr Brendan Wallace is a Research Fellow in the Centre for Applied Social Psychology (CASP) at the University of Strathclyde, with a special interest in the psychology of arousal and attention, systems theory and Continental philosophy. He is also interested in new approaches in psychology (such as ecological psychology and situated cognition). He has been closely involved in the development of taxonomies trialled in the nuclear and rail industries.

Dr Linda Wright is a researcher at the Technical University of Delft in the Netherlands. She was central to the development of the CIRAS system. She has a particular interest in minor-event reporting systems and the predictive power of near miss reports with respect to major accidents and incidents.

Acknowledgements

The authors would like to thank firstly all those people who took part in the studies and projects described in this book for their good humour, support and eager compliance. Their input was fundamental to the development of the ideas outlined in this text.

We are particularly grateful to all those companies in the U.K. rail industry, and the employees of those companies from frontline staff to management, who participate in the CIRAS system. A number of individuals deserve special mention in that regard for encouragement and support including Peter Summerhayes and George Rankin who enthused us with the notion of confidential reporting on the railways, at an early stage. Within specific companies, Robert Plant and Donald McPherson at ScotRail, David Billmore at Great North Eastern Railway, Peter Bowes at Virgin Trains, Gerry Brady and Ian McCallum at First Engineering, Andy Savage at G T Rail Maintenance, and Frank Chambers and colleagues within Railtrack (Scotland Zone), all made fundamental contributions both to the development process, and to our morale, during the early days of rolling out the first version of the CIRAS system.

Thanks are also due to a number of people for help and advice in setting up the initial procedures. These include Peter Tait and colleagues at CHIRP, and Mike O'Leary at BASIS. Closer to home, for continuing gutsy performance over extended periods of time, Helga Reid, and for loyalty, meticulous attention to detail, and infectious laughter at times of crisis, Eleanor Courtney.

Thanks are also due to Maurice Wilsdon and Graham Arkwright at Railway Safety, who funded (along with the railway companies) the later development of the CIRAS project. Especial thanks are also due to Professor Helen Muir at Cranfield University for advice, guidance and moral support during difficult times.

We are also indebted to a number of people working within various sections of the British nuclear power industry, including Iain Carrick and colleagues, Dave Stimson, Steve Harris and Dave Howie, for tangible support and encouragement in our search for the perfect taxonomy! Mention should also be made of the contributions made to system taxonomies and coding procedures by Mathew White and James Harris, who have both moved on to other things; and to Dr Jim Baxter, also in connection with the nuclear work.

Finally, thanks are due to Graham McLachlan for proof-reading early versions of the text, and to all members of staff at CASP and HFA Ltd for good humour and understanding. The number of people with whom we have had productive contact in connection with the ideas outlined in this book is large, and the names too numerous to mention. If we have omitted anyone who feels particularly aggrieved, then the fault is ours; so thanks anyway.

1   Safety, risk and responsibility

Science and subjectivity

First and foremost, this book is about safety and safety management. It seeks to make the point that a number of key features of effective safety management are subjective in nature, and that the role of 'objective science' in this domain, whilst essential, is also limited. But the book is also definitely about *science*. Unfortunately, the word 'science' is still frequently interpreted as meaning 'science according to Newton'. That is, science is seen as the search for truth by individuals without personal bias or motive, who merely serve as the passive mouthpieces for facts and conclusions which are determined by the universe itself. Only one 'truth' exists, and any other account is wrong. A central feature of this view concerns the nature of 'facts'. Facts, it is felt, are either objective or subjective, and only the former are fixed, certain and reliable. Edwards *et al.* (1995) have summarised this view as follows: 'Science is at its best the selfless and disinterested pursuit of truth.' However, whilst this book is about science, it is not about the kind of science described above.

There are a number of alternative ways of looking at science, and consequently when anyone makes use of that word it is quite in order to ask 'What type of science are you talking about'? 'Science' comes in various guises with quite different implications for the types of investigative methods used and the types of conclusions that these methods give rise to. It could be argued that the view of science briefly characterised above, centred around discoverable truths in a fixed and determined universe, represented the state of the art until the middle of the nineteenth century. That view (devastatingly effective for solving problems of a certain class; that is, problems that are *amenable* to solution via that route) has, however, been progressively undermined by new discoveries giving rise to alternative theories about how the universe works, including the theories of relativity, quantum mechanics, and chaos. A brief outline of these alternative 'sciences', and their implications by analogy for work that involves living, thinking, people, is given in Chapter two. Furthermore, the actual *nature* of scientific progress as a system based on objective observations, theory formation, and theory rejection on the basis of a single refutation (e.g. Popper 1959) has been honestly dismantled by Kuhn (1970) who sees the nature of scientific progress as determined by personal motives and as having things more akin to changes in fashion rather than the resolute and dispassionate pursuit of enlightenment.

One of the bases of empiricism (Ayer 1936), and thus a cornerstone of any kind of science, is reliance on the act of personal observation. The argument here is that our own experience of our fellow human beings tells us that they have motives and opinions, that their views are frequently biased, that they have vested interests which colour what they do and say, that their answers to questions vary according to where they are and who they are with and that their opinions, including the opinions of experts and scientists, frequently disagree. If we have personal insight, we will also be aware that the things we do and say are tempered by our own personal interests and situation, and that when arguing a case we are frequently not so much advancing arguments with an inescapable logic, as thinking of a means to defend a position that we feel we have to defend for reasons that may have nothing to do with scientific investigation. Similarly, when it comes to reviewing incidents or near misses, the process of identifying causes and recommending actions to deal with the consequences is frequently not so much a search for 'truth' as a search for a cause that we are prepared to accept as a cause, and for an action that is affordable and that we are prepared to implement.

These things we can readily observe in ourselves and in others. Since we have direct personal experience on these matters, it is *scientific* to adopt methods which take these observed (empirical) facts into account. By contrast, to cling to a view that sees events, circumstances, causes and consequences as ultimately untransformed by the passage through human minds, and as having an objective reality which is knowable and unaffected by human processes, is blatantly *unscientific* since it flies in the face of our own observations. The belief in a certain universe revealed by motiveless scientists is becoming increasingly absurd, especially in areas such as safety management and accident investigation where the selection and interpretation of 'facts' involves human beings at every twist and turn, with all that that implies. In short, it is scientific to adopt an empirical approach which acknowledges subjectivity, and an act of scientific denial to pretend that such subjectivity has no role to play.

What is needed therefore is a pragmatic approach which acknowledges those things which clearly work within a deterministic framework, but integrates this knowledge with an approach that takes into account the variability and uncertainty that arise whenever human beings are involved. Such an integration is the next logical step in the development of scientific methods for safety management, and it can be argued that the sometimes disappointing results of our endeavours stem from a blinkered approach to human action that rules out a reflexive and contextual approach whenever human beings report things or express opinions, and turns a blind eye to the subjectivity of which we are all, secretly, aware. A truly scientific approach, by contrast, takes such things into account within its methodology; as sources of additional information rather than as error. It remains only to add that the failure to take into account the variability and subjectivity of the raw material that safety managers and others have to deal with leads, we believe, to false conclusions, inappropriate or unnecessary actions, and waste of resources, and perhaps most of all, a failure to capitalise on the potential benefits of the available data, much of which is subjective by its very nature.

The need to be safe

Human beings have a need to be safe. In 1943 Abraham Maslow wrote his much cited paper, 'A Theory of Human Motivation' (Maslow 1943) in which he outlined a number of needs which require to be satisfied if human beings are to work and perform happily and satisfactorily. At the bottom end of the 'needs' continuum he postulated the basic physiological needs (so-called 'drives') which have to be satisfied, the classic examples being hunger, thirst and sex. Once these are satisfied other higher order needs come to the fore, of which the first is the need for self-preservation and avoidance of injury; that is the safety needs. As is well known, the top end of Maslow's hierarchy is the somewhat metaphysical (or at least, difficult to define) need for 'self-actualisation', an existential need for self-fulfilment ('What a man *can* be, he *must* be' writes Maslow). However, our main topic for discussion is safety, and specifically Maslow's suggestion that the safety needs, at some point between having a full stomach and playing the Bruch violin concerto, become the dominant source of motivation. 'They may serve as the almost exclusive organisers of behaviour, recruiting all the capacities of the organism in their service, and we may then fairly describe the whole organism as a safety-seeking mechanism.' So people need safety. Survival is, after all, a basic instinct.

Risk and responsibility

However, general statements often break down at the level of specific instances (see, for example, Cardwell 1971: 56–61) so it is useful to look at this general statement in more detail. Overlooking the philosophical issues which arise when any organism is described in terms of the machine analogy (i.e. describing living things as machines is an act of preference; not in itself an act of 'science') there are everyday observations that reveal that Maslow's suggestion about the role of the safety needs, whilst perhaps true at a general level, requires considerable modification at the level of specifics. For example, most people travelling on buses, trains or aeroplanes implicitly expect those who run the buses and trains or fly the planes to transport them from A to B safely. It is a not-unreasonable expectation that the companies profiting from our decisions to travel will take all due care to manage the risks inherent in these activities (though the elimination of all risks is an impossibility). Should they fail to do so, there is ready resort to the courts of law (both at an individual and public level) whenever this duty is not properly discharged. This is true whether those being transported have the intention of travelling in order to spend time in a softly illuminated concrete blockhouse reading a book, or lying motionless on a beach in Spain, embarking on a trail-cycling holiday in Colorado, attempting a solo ascent of Nanga Parbat in the middle of winter, trying to go round the world attached to a balloon, or jumping off the Empire State Building on the end of a piece of elastic. The train, plane or bus is supposed to be perfectly safe even if they are travelling to some remote spot in order to commit suicide! What is clear is that, even if Maslow's assertion is true at a general or population level it makes little sense ontologically, where individual differences and context both appear to play a major role in what individuals see as acceptable risk.

Risk thus appears to be a personal thing. Different people like to do different things, and to take part in activities with differing degrees of risk. This fact is one of the bases for Adams' book *Risk* (1995) in which he argues convincingly that 'objective' measures of risk (i.e. accident or fatality rates) rather miss the point, and in its place offers the twin ideas of a) the 'risk thermostat': a preferred level of risk which is set by individuals and which defines other levels of risk as being acceptable or unacceptable (too high or too low) to that individual and b) risk compensation: when something is clearly risky, people act in such a way that the risk is reduced, so the 'objective' measure is no longer a true reflection of the danger. Despite some of the furore created by Adams' book, we may note in passing that the 'risk thermostat' idea translates without too much trouble into the organisational context in terms of the ALARP (As Low As Reasonably Practicable) principle, which is an entirely subjective judgement. The word 'reasonably' begs the question of 'Reasonable according to whom?'; and the word 'practicable' implies a moveable feast with different things being possible at different times according to developments in safety research, economic situation, public pressure and political will; all things which help to determine where the organisational 'risk thermostat' is set for any particular industry at any point in time. Furthermore, what is 'practicable' is increasingly defined by public and media reaction to the latest disaster, and the perceived likelihood of litigation, rather than being based on any logical safety-related basis.

However, it is also apparent that the notion of what is acceptable risk varies not only between individual people, but within the same person at different times, according to context, and again, quite independently of objective measures. Furthermore, the differences we observe in risk tolerance are socially rather than scientifically or logically defined. Thus, for example, it is estimated that some 50 million ecstasy tablets are taken by young people every year (Murji 1997), with the risk of death being in the region of one in 6.8 million people. This is rather less than the death rate due to aspirin. Ecstasy is nonetheless usually referred to as 'this lethal drug', 'this child-killing drug', or some similar phrase in the media (Sharkey 1996), and the taking of the drug constitutes a risk that those who govern our society are not willing to take, as reflected in the classification of ecstasy as a class A drug. On the other hand, motor vehicles kill two or three thousand people per annum and car accidents are the major cause of death for young people aged 14 to 24 years. Nonetheless, we do not habitually refer to 'these lethal motor vehicles' or 'these child-killing machines' whose possession and use is a source of pride and status to many car owners. Least of all do we indict car salesmen for 'possession with intent to supply'. It is clear that objective measures of risk often bear no relation at all to political and public reactions to societal sources of risk.

Voluntary and involuntary action

A major determinant of reactions to risk appears to concern implicit ethical or moral assumptions about whether a particular risk is acceptable or not. At first sight it looks as though exposure to risk on an individual voluntary basis is viewed differently from risk undertaken on a non-voluntary basis. Thus, we expose ourselves to the risks of car driving on the basic assumption that in so doing we are voluntarily exposing ourselves to certain risks, which (we believe) we know about and understand, that we do so by an exercise of choice or free will, and that we have the skills and abilities to deal with those risks to a level that suits us (the 'risk thermostat' idea). Our fate *seems* to be, and *feels as though* it is, entirely in our own hands. This may be a misperception of course; but a misperception is still a perception. Adams (op. cit.) cites studies from Fischoff *et al.* (1981) indicating that people are prepared to accept far higher levels of risk 'from activities that are voluntary' (we note in passing that this is not always an easy distinction to make, and appears to be more an act of social/cultural categorisation than a logical distinction) and notes that the public is willing to accept risks from activities such as skiing that are several thousand times greater than those it will tolerate 'from involuntary activities that provide the same level of benefit'. However, once again this apparently clear-cut principle receives a spanner in its works when Adams also notes (p. 66) that 'the greater the relative size of the person or agency imposing the risk, the less voluntary the risk will appear to those imposed upon'. In other words, the bigger somebody or some organisation is, the more we like to see ourselves as involuntary victims, regardless of the circumstances, and independent of any logical taxonomy for voluntary and involuntary behaviour.

Thus, when we board a train or aircraft, or go to a hospital, we place our fate in the hands of others, see our actions as non-voluntary, and therefore transfer responsibility to others. Since responsibility implies both the assumption that those concerned had the capacity to behave otherwise than they actually did, and in principle could have chosen *not* to have carried out any act they actually carried out, the failure to ensure anything other than perfect safety leads to the attribution of guilt whenever these two prior conditions are met, or are assumed to have been met (Greve 2001a). However, there are two problems here. The first involves the problem of how to demonstrate that a person could have behaved otherwise than they did; the second derives from the lack of logical clarity separating actions that are voluntary from those that are not.

The actual demonstration that a person could not have behaved different-ly, or could not have refrained from carrying out an action (i.e. that they were not responsible for their actions) is normally tackled from a deterministic perspective, the essence of which is that external (situational/environmental) or internal (neurological) factors were of such a nature that no other course of action was possible, and that the action in question was determined by such factors. This is an approach which is fraught with philosophical difficulties not the least of which is that the issue arises as to whether, or in what form, self determination or 'free will' exists. This will be addressed briefly in a later section. Suffice to say at this point that a deterministic explanation (i.e. a neurological substrate, or environmental/situational demand characteristics) can be found for everything that anyone does if one chooses to look for it (every act has both a neurological substrate or 'mechanism', and every act is situated and takes place in a context) and therefore deterministic science can not justifiably differentiate between 'willed' and non 'willed' actions on that basis. Greve (2001b) tackles this problem by suggesting that whatever the neurological or environmental correlates, conceptually it makes more sense to suggest that willed human action concerns 'things that people do rather than things that happen to them; and furthermore things that they do as distinct from things that their brains do.'

With respect to attributing responsibility, this formulation looks fine semantically, but once again it does not survive a simple test in the real world. If a train driver crashes due to a road vehicle plummeting on to the line just a few seconds before the train arrived at that location at full speed, most people would be content with the conclusion that the incident 'happened to' the driver, on the basis that there was nothing he could do to change the consequences materially. Similarly, if he/she makes a mistake due to an unexpected cerebral haemorrhage we might argue that that is something his brain did and therefore was not 'an action' for which the individual can be held responsible. But what about going through a red light, perhaps at a location where no previous red has been encountered, due to a failure of attention? Do failures of attention 'happen' to people, or do people 'do' them? Is a failure of attention an action or a happening? This can be argued both ways with some conviction in many (most?) cases. Thus we can say with perfect equanimity 'My attention wandered,' as though attention was some sort of strolling vagabond that can quit the fleshly accommodation whenever *it* decides to do so independently of the intentions of the occupier; but we also say 'Pay attention,' implying that attention can be directed and controlled when we wish to direct and control it, in a way that can only be termed 'voluntary'.

At this point the distinction between risks that people take voluntarily, as opposed to risks that are imposed upon them, becomes critical in deciding whether there is blame (guilt) attached to an action, but as noted above the distinction is by no means clear. Most things can be described as voluntary actions or determined actions, the distinction seeming to rest principally in the form of language one prefers to use in describing what happened. Consequently, the idea that this distinction offers a solution to the riddle of why people attribute blame and guilt for misadventure in some circumstances but not in others does not appear amenable to any single, simple empirical or logical formulation. The whole thing starts to look like an exercise in attribution theory (see Chapter seven) rather than an attempt to establish the 'facts'.

In terms of Greve's (2001b) formulation, people choose whether or not to travel on a train, and whilst circumstances may conspire to make this more or less desirable, the decision to use the train is nonetheless an individual decision. People 'do' getting on a train; travelling by train does not just 'happen' to people except in the most extreme and terrible circumstances. The same applies to travel by any other means, or even to the decision to seek medical treatment. (The idea that serious illness 'forces' people into certain types of treatment is not sustainable. Whilst in many cases there may seem no other reasonable choice to make, it is still clearly a choice. Many people, for instance, refuse chemotherapy for cancer, preferring instead a palliative, non-curative course which offers shorter, but better quality, of life). Nonetheless, whilst people choose to expose themselves to known risks on trains, planes, and so forth, responsibility and guilt are still attributed to others when things go wrong.

One of the more vivid illustrations of this fact, that coherent accounts can be given of most behaviours from either a volitional or determinist standpoint, is illustrated in recent controversies about smoking and nicotine addiction. The text *Nicotine Addiction in Britain* (Royal College of Physicians 2000) summarises the major arguments for the proposition that smoking is best thought of as addiction to nicotine; if you smoke cigarettes, 'addiction' happens to you. On the other hand the text *A Critique of Nicotine Addiction* (Frenk and Dar 2000) seeks to disassemble that proposition; and the title of the text *Addiction is a Choice* (Schaler 2000) is self-explanatory.

A cynical, but possibly understandable, conclusion is that no logic underlies the decision to blame somebody for something other than the fact that there is someone or some organisation who can feasibly be blamed. Despite the logical arguments and the research evidence, we seem to be moving to a situation in which responsibility and guilt are attributed when the opportunity arises to do so. This makes things very difficult, because the avoidance of litigation becomes a primary driving principle behind any enterprise that attempts to offer a service to the public, threatening to displace the honest desire to get things done to the best of one's ability. As an example, interviews with managers concerning their motives for adopting an alcohol policy as part of their management strategy cited 'avoidance of litigation' (Davies *et al.* 1997) as their primary reason. This may well not be an isolated example. It only remains to add that avoiding litigation and achieving high levels of safety are not necessarily the same thing.

No firm conclusion seems possible with respect to when people will, or will not, attribute responsibility and guilt; nor in many cases is there any clear logic to differentiate actions that are seen as voluntary from those that are not. We have noted above that the size of the organisation concerned appears to have much to do with whether people will see their own actions as voluntary or involuntary when they expose themselves, or are exposed to, the risks it creates. It seems to be a moveable feast, with differences not only between individuals, but with individuals varying in their willingness to make this attribution according to context. A person whose main hobby is acting as navigator to a rally driver will expect to be perfectly safe in a taxi; so that whilst he likes to take risks in cars, it depends on what car, who is driving, and why. And if, through a lapse of attention, both drivers involve him in a crash, he will hold one driver to be responsible and guilty but not the other one, for reasons that are extrinsic to any logic of action. Individual risk thermostats are thus set at markedly different levels in different circumstances. This is in part, it seems, something to do with whether an act is undertaken, or rather is seen as being undertaken, voluntarily; whether the attribution of blame seems culturally appropriate in that context; but it also has something to do with whether the *opportunity* exists to attribute responsibility and blame, and thus for litigation against another party. Indeed, it could be argued that in some cases the decision about whether an organisation or individual is meaningfully 'sue-able' *precedes* any conclusion about whether one undertook the associated risks voluntarily or involuntarily, thus destroying any 'logical' cognitive model which sees legal action as proceeding from a prior decision about whether an act was voluntary or not. Maslow's 'safety need' therefore seems to have something to do with basic biological drives, but also involves less tangible and more variable components like free will, personal choice, subjective evaluations and perceived economic advantage.

Safety and trust in organisations

There is of course an opposite side to this coin. People offering a service knowing that there are certain risks attendant on the offering of that service have a moral obligation to manage those risks, and to make the risks known to those who avail themselves of the service. The word here is 'manage' however, and that is not the same as 'eliminate'. Elimination of all risks is, as we shall see later, not a feasible enterprise. Furthermore, management of risk demands resources, and the safety of any venture can in principle be pursued to the point where expenditure on risk reduction reaches a magnitude such that the cost of ensuring a particular level of safety makes the service uneconomical to run. Safety costs money, and if it is the case that levels of safety are sought such that the general public (or whoever) are not willing to pay that price for the service, then it cannot be offered with that level of safety. This apparently straightforward argument, that the general public gets the level of safety it is willing to pay for, becomes contentious however, when the situation becomes uneconomical not because there is insufficient cash to pay the wages of those who offer the service, but because the share dividends of people who have nothing to do with running the service are threatened. This then becomes a political, economic and polemical issue beyond the bounds of this text. It remains simply to point out that the combination of high profits/share values and a poor safety record is not viewed favourably by the public and the media, and for some organisations has proved a fatally unstable combination.

In general terms, safety requires that organisations manage that which can be managed, and control that which is controllable (Groeneweg 1996). It also seems ethically desirable to make it plain to the public where there are any gaps in this process. In some circumstances this is fairly straightforward. For example, climbing is an inherently risky sport. Companies like 'Cliffhanger' offer indoor climbing facilities where climbers can practice and improve their skills. Before using the indoor walls, all climbers must have passed a proficiency test in belaying, tying knots and using the right equipment, and the dangers of incorrect use in terms of injury or fatality are clearly spelled out in writing. (It is worth stating, however, that most users of the service know of the dangers of falling from great heights before they go, without the need to have this explained. It seems incredible that anyone engaged in this activity would be unaware of this fact, and slightly absurd therefore that it should *need* to be explained for legal reasons.) Participants also sign a document indicating that they are aware of these risks, and accepting responsibility for exposing themselves to risk (the legal status of this document is not relevant to this argument). But is such an open approach, in which the possibility of death or injury is explicitly spelled out, feasible where a service has a less specialised, less recreational and more general application? Or a service where the users are not themselves *expert* in the risks involved? Does being an expert lead to the personal conclusion that one is therefore more individually responsible for one's safety simply because one knows more about the activity?

This apparently simple 'be honest about the risks' approach is not at all simple when applied to more general public service provision. First of all, the probability of certain events occurring ranges from fairly certain (objects will fall downwards given the opportunity to do so) to extremely remote, and it is impossible for a company to provide a 'comprehensive list' of all the things that *could* go wrong, since such a list is infinitely long. One can compile an endless list of circumstances or combinations of circumstances that *might* occur, even though most of them probably never will. Indeed, estimating the likelihood that certain events will occur on the basis of past occurrences (the basis for probabilistic risk assessment), when such events have occurred very rarely or have not occurred at all in the past, is at best difficult and at worst no better than guesswork.

On the other hand, if there are catastrophic high probability events involved in providing a service to the general public, and these are known to the service provider, clearly the public should be told; but the situation is to all events and purposes basically an empty set since the public are then unlikely to use such a service and therefore such a service cannot be viably provided.[1] The argument only makes any sense in those disastrous situations where such high probability negative outcomes are anticipated neither by the provider nor by the service user. Such examples would include the Thalidomide disaster, the problems with the early versions of the De Havilland Comet, and any other circumstance in which a service or system failed due to inherent flaws which were recognised neither by the user nor by the provider. The provider may then be held responsible, of course, for *not* knowing or for not having tested the product or system adequately beforehand. Notwithstanding, whilst the general proposition that the 'public should be fully informed of the risks' sounds fine in principle, in practice this represents an extremely difficult task. What should a member of the public be told before boarding a train? Should they be told that sometimes trains can crash, a fact of which they are already aware? (Interestingly this is not as absurd as it sounds. Litigation against tobacco companies in the U.S. hinges around the claim that the smokers concerned were unaware of the link between smoking and impaired health, and should have been told of the link by the manufacturers.) Would an exact probability figure indicating the chances that they will survive the journey be meaningful to them, given the subjective nature of risk perception? Should they be told that there is a slight probability that the train might collide with a herd of escaped elephants? Or, more pointedly, should workers in the WTC towers have been warned before the event that there was a possibility that terrorists might, just conceivably, fly airliners into them? What level of information is appropriate between the blatantly obvious, the highly specific, and the extremely unlikely? We note in passing that the number of things that are unlikely must logically exceed that of the likely, and may indeed be infinite; that good safety tends reasonably enough to concentrate more on causes that are likely rather than things that are remote possibilities; and that therefore when things go wrong it should, in an 'ideal world' be due to unlikely things rather than likely ones. The unlikely things, of course, are ones like the elephants on the line that we cannot realistically or sensibly warn people about, other than with some banal message such as 'Beware the unlikely.'

Warning the public that any accident that happens is, due to our excellent safety record, likely to be the result of combinations of circumstances we haven't thought of, is a strange communication to say the least.

Whilst keeping the public 'fully informed' thus seems practically and ethically desirable, in realistic terms this means swamping them with information, some of which will be fatuously obvious, some incomprehensibly technical, and some so remote as to be absurd. The issue then devolves down to one of deciding what to tell them from amongst the wealth of information available; that again is a matter of subjective judgement and selection that involves far more than just objective facts, and insofar as it is selective it involves 'spin'. Suffice to say, what an organisation chooses to tell the public, and elects not to tell them, are the building blocks of the credibility and trust that the public places or fails to place in that organisation. According to some workers, this is where the problem lies for many organisations. However, there is an important qualifier to this statement, namely that the trust that the public puts in an organisation and its communications is heavily dependent upon the organisation's past recent safety performance and competence (Barber 1983). Common sense suggests, therefore, that a poor combination would be a glowing PR message and a poor safety record, and that in such circumstances either the PR message should be brought more in line with observed reality, which could involve grasping a few nettles, or the safety record should be improved if public trust is to be maintained or regained. From this standpoint, a poor safety record is not compensated for by positive spin, but is made worse in the eyes of the public whose trust in the organisation is simply eroded further.

Whilst it may have been the case at one time that both the public perception of risk, and the degree of trust placed in organisations by the public, were simply ill-informed, naïve and irrational (Shrader-Frechette 1998: 45), contemporary work sees the public perception of risk as being as valid as expert evaluations, though based on different criteria. Expert opinion varies considerably, so the idea that the expert view is both unique and 'true' in a way that the judgements of the lay public are not, is not sustainable. Perhaps the expert simply knows more about the technical details; but that does not mean his/her opinions about the implications, *qua* opinions, are any less motivated or biased than anyone else's. Barber (cited in Shrader-Frechette op. cit.: 15) indeed suggests that the lay public are quite right in not trusting organisations when they behave incompetently, and sees such mistrust as an important component of the democratic process, 'a way of maintaining democratic control of authority'.

Equally important, is the fact that mistrust of an organisation by the public, however founded, can materially affect what the organisation must, must not, can, or cannot, do in the future. The importance of the issue is revealed in a number of papers, including Renn *et al.* (1998) who describe the negotiations taking place over the siting of a Swiss landfill site; Lofstedt and Renn (1998) who describe the Brent Spar saga; and Slovic (1998) who describes the 'crisis of confidence' in the U.S. Department of Energy over plans for nuclear waste disposal. Furthermore, trust and competence appear closely related, as do lack of trust and lack of competence. Unfortunately this is not a system characterised by equality. Slovic (op. cit.: 184) argues convincingly that trust is fragile. A single mishap or mistake can destroy public trust in an organisation, and once that trust is lost it can take a very long time to rebuild. Furthermore, an organisation that makes mistakes that it could have avoided forfeits public trust, and in the end may find its destiny taken out of its own hands.

Certain conclusions are possible at this point. We have argued in the first paragraphs of this chapter in favour of that body of opinion that proposes a social definition of risk, and cited some of the strong evidence in favour of that view. Secondly, we have considered the issue of responsibility, and discussed the way in which responsibility, and hence guilt when things go wrong, are attributed. If these arguments hold water, it is clear that companies need to manage risk in two ways. The first of these is the 'scientific' approach based on fatality or accident rates, perhaps taking into account objective results from investigations and studies, and frequently endeavouring to find robust technological solutions to problems. It must also be clear, however, that public perceptions of risk frequently derive from different sources of a cultural and even a personal nature, and make use of different criteria; and also that public perceptions of risk are frequently the prime movers in instigating change in the way that organisations operate. Failure to manage publicly perceived risks (socially constructed risks) can have catastrophic repercussions if a company takes the view that such views are ill-informed, naïve and ignorable.

For example, lack of trust in a railway infrastructure network can result in actions that are immensely costly and yet make only a peripheral impact on the 'objective statistics'. Thus whilst objectively the U.K. railways are still one of the safest forms of transport in terms of journeys commenced and terminated successfully, public perception of the network is characterised by lack of trust and a belief that the risks of rail travel require addressing in a fundamental way, despite the 'objective' facts. Firms thus require to address lay definitions of risk as well as those that are assumed to have a more 'objective' basis, and have only themselves to blame if they do not.

It is often the case that addressing a publicly voiced concern about a state of affairs which is viewed as 'risky' can be done quite simply and cheaply through a more-or-less obvious human-factors (in a general sense) channel instead of through an expensive technological fix. Thus for example, the failure to replace nuts on the tie-bars of points at a rail junction can be addressed, if the problem is known, through an improved monitoring and checking system or through basic redesigning of the points. One of these is cheap and unglamorous; the other has the aura of technological innovation and is very expensive, especially if the new design then has to be back fitted. Or suppose polythene bags of low level contaminated waste fall from a lorry and contaminate a road. Whilst the contamination is slight and easily cleared up, public reaction to the word 'contamination' may force the company concerned into redesigning the polythene bags and the lorry. On the other hand, if someone had reported that in their opinion too much was in the bags, and too many bags were on the lorry, a new procedure and monitoring system would have solved the problem at a fraction of the expense.

Perhaps one of the most striking examples concerns the *Herald of Free Enterprise*, where the failure to fit a bow-door warning light coupled with a delay in sailing, and all that that entailed, led to a public outcry resulting in the need to redesign roll-on/roll-off ferries. The issue, it should be noted, is not whether or not the original design was safe or not safe. No piece of technology is 'perfectly safe', and any piece of technology has to be operated within its design limitations. Whether roll-on/roll-off ferries are intrinsically safe or unsafe is not a matter that the authors of this book are competent to address. The point being made is that the disaster occurred because the vessel was operated in a way in which it was not supposed to be operated (i.e. setting sail, trimmed bow down, and with the bow doors open); and that operating it properly does not cost much more than operating it improperly. Had the incident not occurred, the issue of whether such ferries are safe or not would probably never have arisen in such a stark form. Perhaps the design of ferries would have developed via a more considered process of technological evolution rather than as a forced act of propitiation caused by the failure to operate the existing ones within their limits.

It remains only to conclude that socially constructed risk is accessed generally through what people say, and that the views of non-experts with respect to what is risky and what is not are often ignored, with possibly profound consequences. The second conclusion is that whilst techniques and technologies abound for the assessment of objective risk, and for the identification of error-promoting conditions in terms of ergonomics and design, similar technologies for analysing natural discourse and spontaneous utterances are seldom employed in the commercial setting. In the place of natural discourse we tend to rely on forced choice questionnaires, summaries (interpretations) of the meaning of what was said by supervisors or third parties, or brief descriptions of 'what happened', written in a space barely large enough to hold a name and address by people who are better at talking than writing. Obtaining the maximum benefit from what people tell us is still an art form in its infancy.

Safety culture

One of the ways in which companies seek to erect defences against human error is through the promotion of a 'safety culture'. Many companies seek to monitor safety culture, normally through the use of questionnaires which are filled in by staff. Their responses are assumed to 'measure' safety culture; an assumption based on certain definitions of safety culture which are

in circulation. The questionnaires 'measure' the components of the definition, though in passing it is worthwhile considering the measurement
properties of numerical scales which fail to specify any units (i.e. marking
an answer on a one-to-five scale begs the question 'One to five *what?*').
Some definitions of safety culture are extremely lengthy and complex, making reference to the pattern of beliefs, attitudes, norms, and other intra-
psychic entities which are assumed to have a determining influence upon
what people do, despite the ambiguous evidence for any strong link
between such entities and actual behaviour. The public health literature in
particular is liberally peppered with attempts to change measured attitudes
which, even if successful, have had no impact on subsequent behaviour (see
for example Eiser *et al.*'s 1978 discussion of 'dissonant smokers'; smokers
who accept the health risks but continue to smoke). On the other hand,
some definitions of safety culture are satisfyingly short, such as Reason's
well-known postulate 'The way we do things round here'. Reason's definition has been criticised for its lack of detail and specific meaning, but the
argument here is that the detail in the more complex definitions is both spurious and part of a piece of circular logic. A variant of a quote from Boring
(1923) may be appropriate at this point. He defined intelligence as 'that
which intelligence tests measure'. In the same way, we might define safety
culture as 'that which tests of safety culture measure'; and since the tests
focus on the epistemologically various and uncertain components in the definitions, we have a circular system with no external referent. It also admirably
points out the problems with assuming that changes to the answers one gets
might indicate an improved safety culture, an assumption based on the
belief that something called 'safety culture' exists as an entity outside the
questionnaire itself.

Meanwhile, the term 'safety culture' appears to be bandied about and is
a common figure of speech in many firms and organisations with scant reference to, or even lack of knowledge of, any of its various definitions.
Conceptually, it appears to be an entity independent of any actual person or
act, more like a sort of Calvinistic vapour which permeates some work-
places and not others.

Forced to choose, the authors of this text prefer the succinct version,
'The way we do things round here,' simply because in the best behaviourist
tradition it emphasises observable behaviours rather than intra-psychic
imponderables like attitudes and beliefs which are unobservable directly,
but are naïvely assumed to be isomorphic with what people say in answer to
questionnaires. It suggests that safety culture is what people actually do,
and we can at least go and look at that. The epistemological problems with
assuming that what people say offers a direct window into the way their
brains work is discussed in Chapter seven of this text. Meanwhile, for those
interested in pursuing this matter further, an excellent text by Schumann
and Presser (1996) explores the relationship between the types of questions
one asks, the way that one asks them, and the types of answers one obtains

– but that is a rather different issue. Amongst a number of basic ruses, they point out that using rating scales with an even number of boxes forces respondents to come down on one side or the other of an issue even if they could not care less!

Since we tend towards a position in which speech acts are seen as acts in themselves rather than merely a blurred window into something else (see Chapter seven) we would seek to recast Reason's definition slightly, and suggest that safety culture is 'The sum total of what people *do and say* round here'. From such a standpoint, improving safety and safety culture requires a detailed focus on the acts that people perform and an equally detailed focus on what people say. And what we mean by 'What people say' is their natural discourse; the way they normally talk to each other; and not merely *what* they say, but the *way* they say it. We do *not* mean the verbal utterances (written or spoken) that they produce in response to forced choice questionnaires or surveys, but the way they talk and what they say on their own terms, rather than on terms specified by a questionnaire or researcher. Meanwhile, we suspect that the person who fills in a safety questionnaire in a way that would do credit to St Christopher, before leaving the room and muttering 'What a load of rubbish', is not that difficult to find.

Better value from safety data in a world of diminishing returns

The history of the Industrial Revolution from 1815 onwards shows that in the early days of major high-risk industries, lives were frequently saved, and illnesses/injuries avoided, by a single, sometimes relatively simple and sometimes highly ingenious, fix. Examples that spring to mind at a technical level include the Miners' Safety Lamp, which did a rather better job than the too-much-and-too-late response characteristic of canaries, the introduction of signalling, interlocking of signals and points, and token systems pioneered on the railways and the introduction of the Plimsol line in shipping. In terms of more general industrial health and disease, the abolition of child labour in mines and factories, the use of protective clothing for women working in match factories, the greater awareness of the impact of air-borne particles in the mining and textile industries and the introduction of protective measures, and the introduction of proper sewerage systems in overcrowded cities, are further examples.

The point being tentatively proposed is that, in Western societies at least, many if not most of the more obvious sources of risk and accident have been identified and tackled to a degree. Given the possibility that the most major and most obvious causes bring themselves to our notice soonest, we can hypothesise a process whereby, over time, that which is most appalling and most obvious tends to be dealt with soonest, and the less obvious and less appalling takes longer to become apparent and to address. Progressively, the nature and the causes of subsequent problems become successively less obvious, more likely to be complex, involving a variety of factors, and therefore require a more subtle and sophisticated approach to accident prevention. It also seems reasonable to propose that as major sources of fatality and injury are progressively removed, expenditure on safety suffers from something akin to a law of diminishing returns. To put this another way, in real monetary terms, as society gets safer and fatalities and injuries generally decline, safety becomes more expensive in terms of fatalities/injuries avoided. Notwithstanding arguments about the morality or otherwise of valuing human life in monetary terms, we can probably assert that safety management will eventually reach (or may have already reached in some sectors) the point where the cost of the next safety fix in terms of fatalities or injuries avoided makes the fix unrealistic economically. The preference of the U.K. railways for TPWS (Train Protection and Warning System) as opposed to ATP (Automatic Train Protection) is by all accounts exactly a response to this state of affairs. The cost of ATP per fatality avoided simply ruled the system out of court.

Meanwhile, companies collect data on accidents, incidents, near misses and so forth, which are seldom utilised to their full capacity. Filing cabinets fill with individual reports from which no general lessons are learned and from which at best only a plethora of specific fixes result, with no thought to underlying factors or deep structures which can lead to different surface incidents and issues, nor to the fact that two similar surface occurrences might have totally different aetiology. Furthermore, when something major happens, a full-scale investigation is launched, and because of its intensive nature more so-called 'root-causes' are found, not because they are 'there' in some important way, but simply because we spend more time looking for them. Thus major accidents always look more complex than near misses, when in fact both may be equally complex or simple in substantive terms. Meanwhile, verbal reports of accidents and incidents are mistrusted and characterised as 'merely subjective', and so the collation and analysis of natural accounts and reports is therefore given lower priority than the collection of 'objective facts'. This despite the fact that these latter are also usually communicated verbally by written or spoken word. In this way a major source of information (both explicit in terms of content and implicit in terms of manner of communication) is ignored.

The basic message of this book is that the whole area of risk and safety is grounded in individual differences and subjectivity; that such subjectivity is a useful source of information; that such information can be transformed into useable and reliable data; and that such data are essential to tackling the increasingly complex nature of safety management in the twenty-first century. The information is available through the natural speech, accounts, reports and discourses of people directly and indirectly involved, though its potential is still not realised. Finally, the understanding and exploitation of what lies in people's own accounts is a natural and increasingly necessary complement to the better-developed technological/engineering approach to safety. Quite often a simple and cheap human factors fix will do the job before an expensive and technologically complex solution becomes necessary. The way people see their jobs, what they perceive as risky, what they see as voluntary or coerced action and how they relate to each other may provide important clues as to the most appropriate actions to take; these clues are in their natural discourses, if only we can find them.

Where is risk situated?

As a final note to this section, it is well worth referring to a number of studies cited in the excellent text *Target Risk* (Wilde 1994). Wilde provides findings from a number of simple but classic studies of car drivers, all of which provided evidence of a similar nature. In a variety of ways, car drivers were monitored as they drove over routes on which a variety of data were already available, including the accident rate and the average traffic speed at various locations along the route. The data suggest that drivers experienced more anxiety at spots that were dangerous; furthermore they also adjusted their driving according to the situation, slowing down at locations where accident rates were high, and speeding up where the rate was low. The drivers were thus sensitive to conditions in which accidents tended to happen, experienced more fear, and took more care. The studies provide convincing support for the 'risk compensation' hypothesis referred to previously (Adams 1995). Wilde's conclusions included the following: 'Accordingly, the prospect for greater public safety is unlikely to be found in a 'technological fix' because of the way people respond to such fixes. Instead, the prospect for safety is inside the human being, not in the human-made machine or human-made physical environment'.

Like most good conclusions, the case is probably overstated. There are clearly points in processes where a good technological fix is both desirable and necessary. Otherwise, we are faced with a *reductio ad absurdum* whereby it doesn't matter how dangerous something is, people will always figure out some way of making it safe. That is clearly a ridiculous proposition. Nonetheless, Wilde does bring to prominence an important fact, namely that the human aspects of safety are often ignored in places where a relatively simple and cheap fix aimed at people might do the same, or a better, job than another expensive piece of kit. If that is the case, one way to find out what people think and feel about situations is to listen to what they have to say. Unfortunately, methods for dealing with natural verbal utterances in the safety context remain under-developed. Given the increasing cost of technological fixes, and the decreasing returns in terms of accidents and fatalities avoided, the time is ripe for the development of methods that can capitalise on the information available in natural discourse, in a scientific, principled and replicable way.

2 Safety, subjectivity and imagination

Whilst there are many systems for tackling risk in the broadest sense, they tend to come from the same philosophical direction, and are underlaid by the same deterministic assumptions about people, things, and the way the world works. There is clearly plenty of opportunity for new thinking and innovation both in safety management systems, and in the way in which human factors are conceptualised within such systems. In the meantime, it is perhaps fair to say that some of the actions recommended sometimes seem to go no further than basic common sense. Not that there is anything wrong with common sense, but it is probably not necessary to pay large sums of money to come up with notices to pin on the wall saying 'Think before you act', 'Now check what you just did', or 'Stop, think, act, evaluate'[1] or whatever. These are all eminently sensible suggestions, but could be suggested after a little thought by any sensible person with no claim to any specific expertise.

By contrast, genuinely new approaches, new insights and exciting alternative ways of viewing these common but difficult and important problems are rather thin on the ground. We particularly identify as a problem area the lack of rigorous methodology for dealing with 'subjective' reports concerning accidents, their causes and their consequences, where the choice appears to be between, on the one hand, a loose and undisciplined qualitative approach, in which extracts from reports are selected apparently whimsically by a third person as being particularly salient on the basis of criteria which remain unspecified and unknown, and on the other an approach which seeks to demonstrate its 'scientific rigour' by simply refusing to accept that such subjective reports might constitute useful data in any form, or even be amenable to principled and replicable analysis.

Within the study of psychology itself, there is still a restless struggle between convenient and 'objective' deterministic paradigms, and more modern conceptions of science which recast ultimate truth as either a delusion (i.e. theories based on the relativity of knowledge), as a misconception about the basic processes of the universe (i.e. theories based on quantum mechanics which recast basic processes as probabilistic) or as deterministic in principle but pragmatically unknowable (i.e. chaos theory, within which knowledge is never sufficiently complete to enable prediction of the output of non-linear systems). The strange thing about relativity, quantum mechanics and chaos is that, in various forms, they all point to the human agent as a determinant of what is found out, a shaper of the form that knowledge takes and of what is believed to be 'true'. The human investigator, and the methods he or she uses, are all seen as part of the total system and as having a direct bearing on the results of a study. The Newtonian brand of determinism, by contrast, imagines a certain universe full of objective facts which are discoverable in principle and true in an ultimate sense, independent of who is seeking those 'facts' or the methods used in the search. Since, within psychology, all these theories of physical matter are applied to human behaviour *by analogy*, it remains a mystery why psychologists in particular should so often prefer the one scientific paradigm that specifically excludes, or seeks to exclude, human agency from its studies as sources of 'error', when theoretical physicists by and large have abandoned such a brand of determinism over a century ago. A similar thought has occurred to Koch, who writes, '... the emerging redefinition of knowledge is already at a phase, in its understanding of the particularities of inquiry, which renders markedly obsolete that view of science still regulative of inquiring practice in psychology' (Koch 1964: 5).

Accordingly, in this book we have attempted to address one area of safety management in which we believe progress can be made, namely the ways 'subjective' data in the form of verbal and written reports are viewed and utilised by high consequence industries. Instead of viewing the process of collecting and analysing safety data as basically a traditional 'scientific' enterprise involving objective assessments of objective data by experts, we have tried to view the whole thing from a different scientific perspective, as basically a social process involving such things as personal beliefs and opinions, functional (as distinct from merely 'truthful') communication, subjective interpretation of incomplete or ambiguous evidence, biased[2] and motivated reports by witnesses and experts, selective attention, blame culture, relations between staff, relations between companies, social perception of acceptable risk, political climate and pressure, economics, profit motive and many other factors. We also believe these things are important features of data sets, that they are amenable to principled and replicable study, and that they offer great promise in terms of safety management

We can describe this more succinctly, in terms of three possible models. The first is the classic view of risk management as a scientific (i.e. deterministic/reductionist) process, whereby *hard factual evidence is interpreted by objective and unbiased experts, who would solve all problems if only other people would tell the truth and stop getting in the way in one way or another*. The approach is basically Newtonian in principle, and whilst the subject matter is somewhat more difficult to deal with than the subject matter of classical physics, the assumption is that there exist physically based systems of cause and effect underlying events which constitute 'the truth' about what happened, and that this unique truth can be discovered by an objective, predictable and disinterested science.

The second model takes into account recent thoughts on the nature of science (e.g. Kuhn 1970) and scientists, and asserts that there is no such thing as the 'objective scientist'. Everyone has their own views, biases and opinions, and whilst the opinions of experts may (we presume) have more practical value than the views of non-experts (Davies 1997: 59–63), they are still views and therefore subjective rather than objective. Thus, in a court of law, the prosecution and the defence both recruit their own expert witnesses who, having reviewed the same evidence, arrive at opposite conclusions. For example, at the time of writing this chapter the newspapers are focusing on reports about nuts/lock nuts missing from a stretcher on a set of rail points at Potters Bar. Expert opinions differ on the root cause of this event, even though the 'objective' evidence is supposedly fixed and there for all to see. Nonetheless, some experts are suggesting that the evidence points to a maintenance failure on the part of the contractor, whilst experts from the contracting side believe the same evidence points to the actions of vandals, or possibly sabotage. A third point of view suggests that the nuts gradually unscrewed themselves and that therefore 'no human factors' were involved (overlooking the fact that railways are designed and built by people; they don't design and build themselves!). At risk of seeming polemical, one might entertain the idea that the public and government would wish to attribute responsibility to the infrastructure contractor for reasons discussed in Chapter one, and would require expert opinion to make the point that the contractor was to blame; that the contractor would require expert opinion to make the case that the company was not responsible for the failure, which was due to the actions of badly motivated others, and finally that people who design and install point work would be delighted to hear expert opinion that a new design of points was required. It is clear that expert opinion is not the dispassionate process we suppose, but very much context dependent. The job of the non-expert jury is then to listen to the evidence (i.e. the verbal discourses) of the experts and decide which expert is the most credible. The second model thus asserts that the process of safety management involves *the subjective interpretation of hard factual evidence by experts with their own individual biases, views and opinions.* This must be the case, of course, otherwise experts would always agree.

The third model looks at the nature of evidence. Most of what we know about the world derives from other people's second-hand reports. We read about things, people tell us things, we see things in the written and electronic media. It is rare indeed for us actually to go and verify with our own senses (the basis of empiricism) the 'facts' that are presented to us. Thus, for most of us, most of the time, the evidence is second hand. Furthermore much of the evidence is presented (written or spoken) from a particular standpoint and interpreted by someone else who also has a point of view. The whole process is thus interpretive, selective (how does a person decide what to put in a report, and what to leave out?), and shot through with subjectivity. This, it must be stressed, does not render the exercise pointless, nor imply that research is not do-able. This is merely the nature of the beast with which we have to deal, and only an ostrich could believe otherwise. Furthermore, where events in the past are concerned, we can never be completely certain that our reconstruction of events is correct. As in a court of law, we can *never* verify with certainty that our theories about what happened are the right ones; all we can say is that our theory is, or is not, compatible with the facts as revealed to us, and in the fullness of time we make a decision about which theory we prefer. But in the absence of time travel, we cannot go back and verify that we have actually got it right and it is unreasonable to assume that we get it right every time.[3] Furthermore, most of 'the facts' that we know do not come from direct personal observation, but from other people's written or verbal accounts which are also biased and selective. Consequently, the third model, and the one we propose to explore in this book, is the notion that safety management involves *the selective and subjective interpretation of selected and subjective evidence.*

Readers should note a most important point, however. We are not arguing that the process is unscientific and therefore whimsical and unprincipled. We are simply saying that if we live in a world which is selective, interpretative and subject to individual bias (as opposed to a fixed and certain world in which absolute facts lead to objective, disinterested and inevitable conclusions on the part of motiveless scientists) then the model of science we use should clearly take these commonplace observations into account instead of pretending they do not exist. There is nothing particularly 'scientific' about ignoring that which is obvious to all of us, and falls directly within our range of personal experience.

Knowledge: objective or subjective?

There is, arguably, no such thing as objective *knowledge*. Whilst the world itself obviously has some sort of objective existence, no two people know exactly the same things about the same thing.

Where data do not exist concerning the probabilities of certain kinds of failures (because they have never occurred), the attempt to prevent certain cataclysmic events from *ever* occurring involves acts of foresight and imagination as much as 'science'. In this type of situation one often relies on the judgements of experts. Judgement is a subjective process. The very reasonable assumption is that the informed best guesses of experts will have greater predictive validity than the guesses of non-experts and that their guesses will be more useful more often (Davies 1997). The less reasonable assumption is that the experts will agree in their estimates, which is frequently not the case. One notes the regularity with which safety factors such as 'one-in-so-many-million' are quoted and marvels at the ability of nature to operate in such tidy multiples. Thus one may be forgiven for at least considering the proposition that where failure rates are not known, the intuitions of someone with hands-on experience saying in his/her own words 'I don't like the look of that' might conceivably have as much value as spuriously precise probability estimates, where these are derived from minimal or non-existent data. However, the prize for getting this process right, by whichever route, is that safety is pro-active; that is, it does not involve learning on the basis of the last disaster.

Where the aim is to prevent the *re-occurrence* of some event, the situation looks a little more rational, with the proviso that the mere fact of attempting to prevent a recurrence means that by definition one failed to prevent a particular situation or failure type in the past. Learning on the basis of the last disaster is gained from a hard school and whilst, historically, major lessons have been learned in this way, the process is costly both in terms of company profits, viability, and frequently human lives. However, on a positive note, learning from previous mistakes appears on the face of it to offer more opportunities for 'scientific' investigation. One can search through the wreckage, look for clues, test similar components to destruction, use existing data on failure rates, carry out forensic examination of residues, and generally bring together the whole panoply of 'objective science'. One can also talk to witnesses and others involved of course, a process which would probably be regarded by many safety practitioners as having a strong subjective component, as opposed to the 'objectivity' of the scientific processes just mentioned.

But how real is this division of the world into objective and subjective knowledge? Suppose we start off with the assertion that a broken rail (for example) is an objective fact. We go and look at it, and there it is. But someone *telling* us they saw a broken rail (i.e. *describing* that fact) reduces the 'fact' to the level of someone else's subjective experience; in some sense an inferior form of knowledge from a traditional 'scientific' standpoint (i.e. the statement might be true or not, it might be biased, mischievously motivated or whatever). This raises the straightforward question, 'For the most part, how do we get to know about broken rails, unless someone tells us?' From the view of traditional empirical science (Ayer 1936), the answer is of course, we go and verify it by looking at it ourselves. We verify it through our own individual sense organs. But how often does this actually happen? We can't all go, and if every one of us had to go and personally verify every 'fact' by personal observation, the world would rapidly come to a stop.

It must be apparent that most of the things we 'know' about events come to us via the reports (written or spoken) of other people rather than by direct personal observation. We 'know' far more about the world than we actually perceive with our own sense organs. Consequently, data about unwanted events, by virtue usually of being someone else's personal reports rather than our own direct observations, are subjective in terms of the above definition. From this it follows that an understanding of the fact that verbal (spoken or written) accounts are not simple sources of 'facts' retrieved in un-edited and un-modified form from some computer-like storage facility in the brain, but rather are changing, dynamic and interpretive, is central to the safety process. Safety reports are motivated; people report things because they want something to happen on the basis of their report, they are not simply dispassionate descriptions. They are intended to be performative; to achieve certain goals in the mind of the reporter. There is, arguably, a presumed (but by no means always justifiable) qualitative dimension to this subjectivity, however. We trust some people's verbal reports, for better or for worse, more than other people's. But we still have to confront the conclusion that, setting aside our own personal direct observations, most of what we know about unwanted incidents comes from other people's reports, from people with differing levels of skill, different biases, favourite theories, differing degrees of knowledge, differing motives and so forth. It is therefore selective and interpretive, that is subjective. Even where our own personal direct observations are concerned, the mere fact of our writing about them, or describing what we (think we) saw to someone else, immediately changes the nature of the information. To the person I communicate with, my direct observations are now merely someone else's subjective report. To report something is to transform and interpret it.

Does this mean therefore that science is not 'do-able' in this area? The answer to this is, quite definitely, 'no'. But it does depend on what kind of science we are talking about. Instead of a science that sees ultimate knowledge as totally objective, the universe itself as in principle determined and predictable, and all the details of the universe as potentially knowable by motiveless scientists, we need to take a look at sciences that take into account the subjectivity inherent in the acts of perception and observation, the fact that observation itself transforms things, and the unpredictability that resides at the heart of everything. It is worth noting that previous authors have given more than a second glance to the possible utility of alternative contemporary scientific paradigms. Thus, Groeneweg (1996: ch.7) gives extensive consideration to the possibility that accident databases may best be described in terms of chaos theory; and in a broader systems-theory context Carver and Scheier (1998) have discussed the applicability of both chaos and catastrophe theory to the study of human behaviour. It may therefore be appropriate at this point to give a short, naïve account of the three paradigms that have replaced Newton in the search to find out what the universe is 'really like'.

What kind of science?

It is worth while examining the proposition that science discovers 'truth'. If that were the case, science would eventually grind to a halt, as there would be nothing left to find out (Lawson and Appignanesi 1989). It is only the fact that science is always wrong or incomplete that allows the process to continue. What science actually does is solve problems ('Under normal conditions the research scientist is not an innovator but a solver of puzzles' [Kuhn 1970]) and as theories develop and change, it solves them in different ways, possibly better ways. But the final 'truth' is probably unknowable. Therefore, the goal of science is probably rather more humble. Its goal is pragmatic. If a theory enables us to solve a problem, we say it is 'true', till a new, different or better theory comes along; and that will always be the case. Truth is merely what works.[4] Thus Gribbin (1995: 230–1) speaking of particle physics, writes 'none of our theories and models provide "the truth" about the particle world, and all of them are more or less successful attempts to provide a picture we can understand and models we can use to make predictions with'. So scientific theories help us to solve problems. Whether they are true or not is not knowable, and furthermore is a red herring. The 'truth' of a scientific theory is the answer to the question, 'Does it work?'

For a great many things, Newton works. However, there is a range of things both within the physical and human domains where Newton does not fit the facts as we understand them. One of these areas is human action. It is interesting to note that the theories of Einstein and Heisenberg, and perhaps to a lesser extent Feigenbaum, seem easily compatible with the idea that the scientist or investigator is a central part of the epistemological process, and that his or her actions are a) a part of the physical process being investigated, and b) determine the nature of the results obtained. Subjectivity thus appears to be an acknowledged component of these theories, with the status of the observer, and the act of observation, having a determining effect on the results. In a sense they place the scientist in the physical world alongside the phenomena investigated. By contrast, when applied to human behaviour, a Newtonian approach increasingly takes on the appearance of the study of the world by a group of super-beings who don't belong to it, who are unaffected by it, and who operate according to different principles.

Relativity, quantum mechanics and chaos

At this point it is worth re-hashing some of the more readily accessible implications of the theories of relativity, quantum mechanics and chaos theory, to see how these offer certain advantages over a Newtonian world view as philosophies for studying human action.

Relativity

Perhaps the single most important implication of the Einstein analogy for the human sciences is the suggestion that, even for physical objects and systems, measurement is relative rather than absolute. The simplest way to illustrate this is by means of the overworked 'train' analogy. Suppose I lean out of the window of a train (a very poor idea) travelling at 80 miles per hour, and throw a ball towards the rear of the train at the same speed. The ball will travel backwards *relative to the train* at 80 m.p.h. However, *relative to the ground* it will go nowhere. It will hit the ground at exactly the point where I released it from the window. To an observer *on the ground*, the ball will simply fall vertically from the point where it left my hand. By contrast, if I throw the ball forward to a friend standing ahead of the train (presumably a friend who doesn't mind being hit by trains) waiting to strike the ball with a tennis racquet, the racquet may well be ripped from his grasp as the ball hits it at 80 m.p.h *plus* the speed of the train i.e. 160 m.p.h. This illustrates a simple fact, namely that the speed of an object, to an observer, depends on the speed and direction of the place from which the object comes; and also, reciprocally, where the observer is and how fast he/she is moving (relative to something else). The speed of an object, therefore, is not an absolute measure. By extension, the theory of relativity implies that measurement is meaningless unless the conditions under which the measurement was taken are specified. That is, unless you answer the question, 'Relative to what?' the measurement is arbitrary.

If we accept the analogy as having more general application, then we need to specify the conditions under which we measure things, and accept that under different conditions the results may be different. Obviously, if we are measuring something physical with a ruler or a micrometer, the differences due to relativity will usually be negligible and we may be able to ignore them for most purposes. However, in the human sciences analogous types of relativity effects are common and of immense importance. The answers one gets to questionnaires or in interviews are seriously affected by the way one asks the questions, where one asks them, why one asks them and so forth. Other things which affect people's answers include the nature of the instructions, the person's beliefs about what the survey or questionnaire is for, what they believe the consequences will be of certain kinds of answers, the way the questions are worded, and whether they want to take part or not. They are even affected by who asks the questions. For example, a group of people interviewed about certain health problems (drug use) in a university office by a clinical researcher gave quite different answers from those they gave when asked the same questions in a place of their own choosing by one of their peers (Davies and Baker 1987). There are numerous examples of these kinds of effects (see Davies 1997: ch.8).

Questionnaires in particular contain the seeds of the answers they produce, sometimes in a very obvious form. By contrast, the demand characteristics of natural discourse are less intrusive (Davies and Best 1996; White and Davies 1998). To put this another way, forced choice inventories, rating scales and questionnaires are good ways of obtaining verbal responses to issues that concern the investigator. Whether people actually think about these things at other times, or possess attitudes to them in any meaningful sense of the word, is a matter for debate. On the other hand, minimally cued natural discourse is a way of finding out what the *subjects* of an investigation think about, as opposed to prompted verbal utterances about the preoccupations of the researcher.

The first proposition, therefore, is that when people answer questions, respond to surveys or fill in questionnaires which measure various attributes, the interpretation of their responses requires that we take into account the circumstances in which the information is obtained, and that we interpret the data within that context. A different approach can and often will produce different results.

Quantum mechanics

A naïve understanding of Heisenberg requires us to accept that the stuff of the universe runs on a probability basis rather than a deterministic basis. At the level of sub-atomic particles, it is impossible to predict the behaviour of the stuff of which the universe (and incidentally, that includes people and people's brains) is made. Instead, one can only make probabilistic statements to the effect that a certain quantity of stuff will do this, and a certain quantity will do that, without being able to say which individual particles will do what.[5] At this level, causal prediction in terms of underlying mechanism disappears ('There are no wheels and gears beneath this analysis of nature,' writes Feynman [1985: 78]). But quantum experiments reveal even stranger effects; effects for which no deterministic explanation exists but which nonetheless can be reliably observed. It is interesting, as a precursor to the next few paragraphs, to take on board the quote from Richard Feynman (in Gribbin 1995: 246), 'Do not keep saying to yourself, if you can possibly avoid it, "But how can it be like that?" ... Nobody knows how it can be like that.'

In classic studies by Nils Bohr (cited from Gribbin op. cit.: 10–19) individual particles (photons) fired through slits in a partition, and thence onto a screen appear to 'know'[6] whether they are being observed. If a particle counter is placed near the slits to count the photons, they (the photons) behave like particles and produce little blobs on the screen. Take the particle counters away, and the photons produce an interference pattern, a property of waves not particles. The photons thus appear to have a remarkable property. They behave differently ('know') when they are being observed and when they are not. In that sense, therefore, they are very much like people, who behave one way when they know they are being observed, and a different way when they are not. But there are even more amazing phenomena to account for. It is only necessary to put a particle counter on *one* of the slits to produce the same effect; the photons going through the other (non-monitored) slit 'know' that the neighbouring slit is being watched! If that is not fantastic enough, the story gets curiouser and curiouser. In studies involving a Pockels cell (a Pockels cell is basically a detector which can be switched on or off very rapidly; the transit time is nine billionths of a second), the cell can be switched on *after* the light has gone through the slits, but *before* it hits the screen. The switching of the Pockels cell was also delegated to a computer which randomly decided whether the detector would be on or off on each trial. Under these circumstances, the photons produce wave patterns when the Pockels cell is off and particle patterns when it is on, even though they have already gone through the slits in the partition, and thus by implication, 'decided' whether to behave like particles or waves (Gribbin op. cit.: 140). To quote from Gribbin, 'The behaviour of the photons ... is changed by how we are *going to* (emphasis in original text) look at them *even when we have not made up our own minds about how we are going to look at them.*' It appears that the behaviour of the photons *now* is fundamentally affected by something that is about to happen in the future. This looks alarmingly like backward causality.

The implications for the quantum view of the world for psychological studies are really rather obvious. Whilst it is not at all clear from quantum theory whether in any literal sense the stuff of the universe 'knows' when we are looking at it, its behaviour is affected by the act of observation; so is the behaviour of people. There is a vast literature on 'social-facilitation effects', 'conformity effects', and 'experimenter artefacts' which illustrates this phenomenon. One of the best known and most pertinent here is the classic series of industrial studies known as the 'Hawthorne Studies' which examined the assumed determinants of productivity at the Western Electric Hawthorne plant in the U.S. (Roethlisberger and Dickson 1939). Originally undertaken to assess the effects of lighting levels, work breaks and other manipulations of working conditions on productivity, the studies found that the supposed 'determinants' were not determinants at all, and that productivity was more a function of informal group organisation and the motivating effect of being part of an important study organised by the Harvard Business School. Some groups managed to maintain their output when virtually working in the dark. The conclusion from the original studies was that output was largely independent of the physical working conditions, and more a function of relationships within the group, with supervisors, and with the fact that the group felt 'special' as a consequence of being observed as part of an important study. Argyle (1972: 106) also refers to 'the general expectation throughout the works that the experiment would be a "success"'. Such effects are commonplace, and the effects of being observed on behaviour have been well documented (see for example McClintock 1972: Section 3). It has even been shown experimentally that such social-facilitation effects can be induced simply by placing a tape-recorder in a room and the effects of cameras on public behaviour is one of the cornerstones of modern policing in high-crime areas. It should be noted that the literature looks at situations in which performance is enhanced by being observed, *and* at situations in which performance deteriorates. It also appears that even cockroaches show these social-facilitation effects when they are watched by other cockroaches (Zajonc *et al.* 1969).

The implications of the quantum theory analogy for the human sciences must surely be obvious. If we simply translate some of the observations made by quantum physicists about the basic matter of the universe into the human domain, they are not even contentious. People know when you are looking at them, assessing or monitoring them, and that knowledge affects what they do and how they do it. Note how the sight of a police-car in the rear-view mirror changes how we drive. The mere *thought* that we are being observed affects behaviour; note the effects of speed camera-boxes, some of which have no camera inside. Also, in a sense, backward causality is commonplace. As a matter of routine our choice of behaviours is determined by what we know is likely to happen next. Finally, in a very real sense, effective safety management is very much shaped by future consequences, involving as it does the need to regulate our behaviour *now* in relation to events in the future.

Chaos

Chaos theory is a mathematical theory. In this text the word chaos refers to that theory; not to the state of Paisley Road shortly after the finish of a soccer match, played at Ibrox, between Glasgow Rangers and Celtic. Chaos theory is less associated with the work of a single person than either relativity or quantum theory, though Feigenbaum is often credited with taking some of the initial steps in this area (Gleick 1998). Like the theories of relativity and quantum mechanics, chaos theory fundamentally challenges certain traditional beliefs in the ability of science to reveal absolute truths about the universe. Chaos theory simply contradicts the idea that if we know enough about a situation *now* we can predict what will happen next, on the basis that in certain kinds of systems one can never know enough *in principle* about the situation.

The basic idea is that there exist non-linear systems which are *sensitive to initial conditions* to such an extent that accurate prediction is never possible. Some of the earliest work on chaos concerned the speculations of biologists about fish populations in ponds, and how these populations would increase or catastrophically decline in quite unpredictable ways as a consequence of the tiniest adjustment to population pressure parameters. More importantly, when the parameter is altered by the tiniest fraction (maybe the tenth decimal place for example) the population either flourishes or becomes extinct. That is, the sums yield completely opposite outcomes as a result of the tiniest change in the parameter, never settling down to steady state, nor to steady growth, nor to steady decline, but oscillating wildly between completely different outcomes in ways that are *in principle* unpredictable.

In other words, when one is measuring things, there is a limit to the degree of precision one can achieve; a limit to accuracy; a limit to the number of decimal places one can handle. But at that point there are always plenty more decimal places potentially available. When a system is sensitive to initial conditions (i.e. *chaotic*) outcomes are never predictable because at some level there is sensitivity to a measurement level that has not been attained. And such levels will always exist. The point about chaotic systems is that these tiny differences in parameters produce massively different outcomes. Whereas within a linear system, a small change in a variable would be expected to produce a small and insignificant change in the result, in a chaotic or non-linear system a small change in one parameter can turn triumph into catastrophe.

The second implication of chaos theory is that even if one attempts to collect information of the finest grain imaginable, the data eventually become *fractal*. Fractal means that, instead of eventually finding some ultimate 'truth', all that one finds is more of the same. The best illustration of the fractal effect is probably the Mandelbrot Set (see Gleick op. cit.: 221) derived from photographs of successively magnified objects. At higher and higher magnifications, all that happens is that patterns seen at *lower* magnifications keep recurring and recurring at higher ones. But the picture always looks basically the same, nothing fresh or new emerges and so one is 'none the wiser'.

The implications, by analogy, for safety management concern the level of detail into which we go when looking for causes in data. Where a system is chaotic (we accept there are problems in determining when a system is in fact sensitive to initial conditions) prediction will never be possible; this is one reason why accidents continue to happen despite our best efforts. The second, somewhat more stretched, analogy suggests that in some cases we may investigate things in greater and greater detail, see more and more, and be none the wiser. For instance, in the wake of a catastrophic event, the natural tendency is to look for the causes in a more intensive and detailed manner than is the case for those events which are less serious. Thus the systems for investigating incidents in the health service proposed by the U.K. National Patient Safety Agency (National Health Service, Scotland 2002) allocates more time/resources to the investigation of Code Reds (events with serious consequences) than to the investigation of less-seriously rated Code Yellows or Greens. This may look logical but it isn't. The consequence is that more root causes will be found for Code Reds since more time is spent looking for them, and thus serious events will always look more complicated than less serious events, without necessarily leading to any practically useful conclusions. For example, a confidential memo (2002; non-attributable) to the author from a U.K. health agency reported that a recent investigation into an incident yielded 179 unprioritised action points. Where serious events are concerned, therefore, their complexity may well be a function of the search strategy we use rather than a simple property of the event. Furthermore, we can end up with a massive list of things to fix with no evidence to assist in deciding on their priority. That is, there are more, but they are all the same. By contrast, near misses will always seem simple.

However, *prima facie* it seems highly unlikely that as a general principle disasters must by their nature always be more complex than near misses, and quite possible that disasters can occur for very simple reasons just as near misses with few observable consequences might have complex aetiologies from which lessons could be learned about preventing disasters.[7] This latter suggestion, of course, implies (as in the 'triangle' models of Heinrich, Bird and others) that the 'causes' of major events are to be found amongst the more numerous minor events (unsafe acts), so that removing the root causes of minor events will also reduce major ones. This is a big 'if' of course, and Chapter 3 in this book discusses the 'triangle' models and addresses that specific issue.

One other implication of chaos theory is equally apparent. Namely, since there is no such thing as perfect knowledge, prediction except in the most trivial sense is also imperfect and consequently, contrary to the mantra of some safety-system developers, zero accidents is a foolish (i.e. non-achievable) goal. In more general terms we can also see chaos theory as implying that a piece by piece reductionist approach to accident/incident investigation superimposed on a belief that the more detail to hand, the better one can control and predict – supported by an apparently common-sense but untested assumption that major accidents should *obviously* be investigated for more detail than minor incidents – could be self-defeating.

Are these things anything more than analogies?

For anyone who finds this type of analogy between contemporary physics and behavioural science incredible, it only remains to add that physicists appear to make the analogy quite readily. Newton's theories, in any case, are also theories of physical matter applied to the study of human action by analogy. If the application of the ideas of Einstein, Heisenberg and Feigenbaum to the realm of human action seems to require too great a leap of faith, we would argue bluntly that it is no more strange than the application of Newtonian physics in those realms, and possibly less strange since these more recent theories do at least acknowledge the role of the observer in physical processes.

Furthermore, the degree of looseness or fuzziness introduced by contemporary views of science appear to be properties of the world, and to have implications for what we regard as useful knowledge. Conceding that such looseness is inherent in the accident/incident investigation process simply brings human factors psychologists into line with contemporary thinking in the physical sciences; the physical sciences being the branch of knowledge that psychologists have, by and large, sought to emulate since the birth of the subject. Bringing a scientific approach to subjective data is, after all, what psychology is all about, and simply designing studies in such a way that subjectivity is ruled out, or not allowed to participate, is nothing less than throwing out the baby with the bathwater.

Causality: a property of the world, or all in the mind?

In the wake of an incident, the basic task of anyone interested in safety management and incident/accident investigation, where the primary goal is to establish a 'root cause' or causes, is to find out what caused what and to seek from amongst the available data the nature of those causes that led to the undesired consequences and in whose absence such consequences would not have occurred. The previous paragraphs have attempted to point out how a number of the components in this task have a strong subjective element, and that scientific paradigms exist which allow for this fact. It seems fitting therefore at this point to review some of the arguments suggesting that even the concept of causality itself has as much to do with the observer as with the state of the world *per se*.

Events, causes and consequences are not simply properties of the physical world. They are percepts. Events, causes and consequences have to be perceivable properties of the world, or we would not know they were there. They have to involve perceivable changes in the state or position of physical matter[8] before we notice them. Events that we do not perceive cannot have causes as far as we are concerned and their consequences would appear to defy analysis. We cannot identify the causes of events we do not perceive, nor plan any corrective action if we do observe their consequences. We would not even realise that they were consequences and their propagation would seem to be magical.

We can fail to perceive an event for various reasons. Most obviously, we may physically not be in the right place to observe it directly or to receive the reports of others. Alternatively, there could be human factors considerations; for example we may not be paying sufficient attention to notice the event; the event may be below our threshold of perception (i.e. too small to see or hear); or it may take place so slowly that we do not notice it (this latter factor is interesting insofar as it implies the possibility that *too frequent monitoring of a slow process may fail to reveal change*, for example deterioration of materials, or erosion of a work practice – it is a continual source of amazement how slowly tomato plants grow, only to shoot up during the week one is on holiday). Whatever the case, the obvious, but crucial point is that from the point of view of the safety engineer, who has to act on what he or she sees, hears or senses in some way, an event is not an abstract occurrence that happens 'out there' regardless of whether anyone sees it or not, but a *percept*. Consequently the thing that is acted upon is not some 'absolute event' whose parameters are clearly defined and incontrovertible, but something that is filtered through a human mind, or possibly several; an expert construction of an event based on what someone sees or hears, what they know about it, and how they interpret the information. It should be noted that no two accounts of an event are ever the same, but vary according to the perceptual and cognitive biases and preferences of those who observe it. However, there are some commonalities in these processes.

In our everyday lives, we tend to group things that happen into clusters of three. Each cluster consists of (a) causes (b) events, and (c) consequences. Typically, the *event* is something that happens *that we notice*. Thereafter, the thing or things that happen just *before* it we tend to call *cause(s)*, and the thing or things that take place *afterwards* we call *consequence(s)*. The event is thus the thing that initially catches our attention, and it can do so for a number of reasons. It is normally conceived of as being brought about by (caused by) things which precede it, and as having an impact on other things afterwards (*consequences*). In terms of simple systems of cause and effect, this represents for many people the way the world *really is* and the way it *really works*. It is proposed to start with this simple model since it serves as the springboard for the arguments that follow, notwithstanding the fact that the assumption that the world really works this way is almost certainly wrong.

The basis for suggesting that this common-sense system is wrong (NB the fact that it is wrong doesn't mean it isn't useful) stems in the first instance from certain arguments about causality put forward by Hume (1748/1962) who argued that causality was 'an union in the imagination'. Such a viewpoint puts human mental activity in primary place as the organiser of chains of causality rather than the physical world. More recently, Lana (1991) has written 'the idea of causation is ... an epistemological attribute of human beings' (42) and 'causation depends on the epistemological characteristics of the inferrer rather than on the characteristic [sic] of the objects seen as cause and effect' (43). However, the route to this general conclusion is rather different in the two cases. Whilst Hume argues in a highly abstract way for the idea of causation as arising from contiguity, succession and constant conjunction, and includes some rather subtle consideration of where impressions come from in the first place (from the object itself, from the perceiver, or from God), the current argument is much more mundane. Our experiences suggest that the division of the world into causes, events and consequences varies with the individual, with the purpose to be served, with the context, and increasingly one suspects with the amount of energy and resources necessary to fix something once it has been identified as a 'cause'.

Let us examine more closely the distinction customarily made between a cause, an event and a consequence. It can be argued quite cogently that an event does not exist in isolation but is itself the product of (i.e. a consequence of) a cause; that is there is no such thing as an *uncaused* event.[9] An event is always a *consequence* of a prior cause, and thus the act of calling it the *event* merely indicates that it is the thing (the *consequence*) that we are most concerned about; but it is always a consequence. However, to be observed, an event must also be a physical change of state of some kind, and any change of state must have implications for other physical matter. Insofar as this is true, an event is also always a *cause*, since its occurrence changes the state of the universe and thus has consequences. This of course is an infinite regress both into the past and into the future from some point in time *now* which specifies the time at which we choose to locate the *event*. Into the past because the cause of the event, in order to be observable, also has to be (or involve) a perceivable state or change of state which must then *itself* have a cause, and into the future because any event is also a cause insofar as it has consequences which cause other things to happen. From this point of view, the trio of 'cause, event and consequence' is revealed as an arbitrary linguistic device for ordering the universe, for helping us to communicate what things we see as important, what things we intend to fix, and what things we wish to avoid, rather than a pronouncement made solely under the aegis of theoretical physics.

Domino theory: the chain of causality. Does it exist?

In terms of everyday thinking, things happen because they have causes. After all, causes are what make things happen. Once they have happened, there are consequences; consequences being basically the aftermath of events happening. Viewing the world this way helps us to make sense of it and to find ways of preventing bad things from happening again by removing causes; the consequences are the ways in which events communicate their positive or negative impact to us and lead us to decisions about whether we wish them to happen again or not. It seems so natural to see things in these terms that to suggest that such 'chains of causality' are at best arbitrary, and at worst illusory, seems the height of nonsense.

However, despite the obvious phenomenological strengths of the above, it is easy to see how arbitrary the premises upon which this repeated cycle of cause, event and consequence rests really are. In order for us to observe that something is a cause, the perceived cause must itself be an observable change of state. Things which are steady state or 'constant' simply do not feature in our explanatory accounts. The body of psychological theory known as *attribution theory* makes this very clear. Thus Kruglanski *et al.* (1983) write, 'if one of the entities covaries with the effect while the remaining ones did not, the covariant entity is the events "cause"'. A more comprehensible way of expressing this is through analogy, and one cited in the literature by a number of authors concerns the question, 'Why did the house burn down?' An emphasis on *constant* factors could produce an explanation in terms of the propensity of oxygen to combine with various materials to produce oxides at high temperatures. But since this is common to *all* fires it has no explanatory value in terms of why this particular fire took place. On the other hand, 'Jack fell asleep in bed whilst smoking' is a non-common factor which hence has explanatory value.

In this regard, therefore, any causal account tends to be built around non-common factors before we regard it as having any pragmatic (useful) value as a satisfactory explanation of why something happened. And since such factors are non-common they must be observable as discriminable stimuli to people in terms of situations, events or changes of state in their own right. In other words, causes themselves have to *happen* (even if this is gradual, over a period of time) in order to be observable. However, if causes happen they are, phenomenologically speaking, events in themselves, and if they are themselves events then they also must have precursors i.e. causes. It is thus apparent that this process is an infinite regress, with every cause having its own precursors which in turn cause *it*. Consequently, there is ultimately only one root cause of everything, namely the big bang, the moment of creation; the only event about which we no have prior information, and consequently lack the raw materials from which to fashion an attribution, other than God.

However, if a cause is also an event, it can also logically be a consequence. We have seen how something we initially described as a cause can be recast as an event simply by focusing our attention on the thing or things that preceded *it*. Given that that is in fact the case, the thing that preceded it (i.e. the 'new' cause) must also have its own precursors and hence be capable of being recast as an event also. From that standpoint, the very thing to which we *originally* ascribed causal status now takes on the status of a consequence, since it comes third in a chain of causality.

Consequently, the division of natural phenomena into neat cycles of cause, event and consequence is revealed as not having any 'tight' or necessary isomorphism with the natural world itself, but as a motivated act of categorisation on the part of human beings who wish to subdivide the world so. This is particularly the case where factors having nothing to do with 'pure science' (fear of litigation, market impact, public perception, economic pressures) enter the equation. Causes, events and consequences are thus not properties of the universe but properties of people; a moveable feast whose starting point is governed not by any physical reality but by the pragmatics of problem solving. The *now* event is selected because it is the thing that strikes us as the most salient from our own subjective viewpoint; the *cause* is so called because it represents the point at which we feel action is necessary, possible and/or affordable; the *consequence* is an entirely evaluative dimension (did we like it or not?) on the basis of which we decide whether we want to change things or not in order to prevent a recurrence. In other words, the tripartite division of the world into causes, events and consequences is a motivated act of construction that helps us deal with things, and which we impose on a continuous and undifferentiated process of 'what led to what' which otherwise goes back to the birth of the universe.

Context and the perception of cause

A man attends a stag party for one of his work-mates, has several drinks, and arrives home late. He turns up the following morning at the power plant where he works, still slightly the worse for wear. Later that day, whilst carrying a tray of tea and biscuits, he falls down a void and breaks his leg because a cover plate was not replaced after maintenance. In order to code this event we have to consider a number of questions, which include the following. What was the event? What was the cause of the event? What were the consequences of the event? Is it more than one event? If so, how many, and what are they? What are the causes and consequences of these events?

A basic problem for any coding system concerns the difficulty of defining 'What is the event?' That is, what is the event for which we seek a cause? In the present example, is the event the act of breaking a leg, of falling down a void, of leaving a cover plate off a void, or of carrying trays of tea about in a dangerous environment? And what is the cause? Is it falling down the void, or is that a consequence of something else? Is the root cause actually the happenstance of having a friend who chose to get married on a particular day? Or of wanting a cup of tea?

More or less out of force of habit, the need to isolate the 'real event' seems paramount because of the appealingly 'logical' way we are taught to think about chains of causality, and such logic is the underpinning for the way science is supposed to proceed. What we decide to call the event determines what we see as the consequences, and locates the causes on which we shall attempt to take action. The answer to the initial question, 'What was *really* the event' is however in this case both unsatisfactory and unanswerable. The event is whatever we want it to be, according to our priorities! But having said that, the problem is far from being insoluble by logical means. As we shall see later, the crucial factor in this type of dilemma, especially where the data are entered onto a company or possibly a national database, is purely and simply reliability of coding. Given that any component in this chain of incidents could be called the 'event', and the lack of any fixed logic for doing so, the one thing that becomes important is not that different coders should all get it 'right' (there is no 'right' answer) but that they all be trained in such a way that they will *agree as to how it should be coded.*

If different engineers code the same sequence of events differently, then differences between data sets and databases represent differences between engineer's coding habits that are not reflected by differences in the real world, and so it is impossible that any lessons can be learned. Paradoxically, however, if they *agree* about how they will code an event (that is they all code it in the same way) we have a tool for acting on the real world. Thus if everyone always agreed that something was 'black' whenever it was 'white', we have a reliable and robust marker against which we can test associations and attempt to make predictions. The problem only arises if some coders say 'black' and some say 'white'.

From this point of view, consensus statements (agreements) can form the raw materials for prediction and prevention, without the need to get into arguments about what the world is 'really' like. Thus, if the plant blows up whenever everyone says 'black', the thing to do is look out for those occasions when everyone says 'black' and concentrate on stopping the plant blowing up, rather than become involved in a philosophical discussion about the objective reality of people's colour judgements.

'Root cause analysis' involves the search for that factor or those factors in the absence of which an 'event' could not have happened; but this again is a motivated rather than an objective process. Clearly, it is only possible to embark on this search *after* the event has been identified. To return to our home-spun example of the stag party, in the absence of such definition, root cause status could be apportioned to a number of variables, including the wife-to-be of the friend whose party the man attended the night before. If we *wanted* to, we could select the event in such a way as to make out a strong case that it was all her fault! If she hadn't agreed to get married on that day, there would have been no party, the power plant worker wouldn't have been hung-over and hence might have been more aware of what he was doing and where he was walking on the day the maintenance fitters left the cover plate off.

At a second level, there are even more problems. By virtue of the work they do, different people will have different priorities and different views about what is or is not salient. In the medical wing, the central event will be the broken leg and all the activity will focus around that. Back in the maintenance workshop, investigations are in progress to find out what caused a cover plate to be left off; *that* is the event and the broken leg is merely a consequence. Meanwhile, the HSE man wants to know why someone was carrying trays of tea about in a high-risk area. As a result, since there are many implicit but probably unstated definitions of the event, there are many different root causes and though they appear to bear on the same issue they are actually root causes for different events. Failure to define the event explicitly thus leads to analyses which produce a plethora of root causes, sometimes hundreds, in which focus is lost and unconsciously the aim gradually transmutes into a desire to fix everything at once. (Where a disaster is involved, this may represent an uneconomical act of public atonement rather than a logical search for the primary causes.) This may be supported by the unstructured use of counterfactual reasoning (arguments of the type 'If this had not happened then that could not have happened') which has the potential to produce a virtually endless but unprioritised list of possible 'root causes'. (For a detailed account of a mathematical approach to counterfactual reasoning, see Johnson 2001.) If finding root causes has a strong subjective component, then some structure has to be imposed on this subjectivity if people's judgements are to prove useful.

Safety and imagination

According to 'scientific' models, the causal chain is ordered on the basis of time. So, an event 'happens'; this is the case because certain conditions preceded it in time and are thus the 'causes' of the event; the 'consequences' are things that come about afterwards as a result of the event happening. Classic regression-based models of the type described by Blalock (1964) basically commence from this starting point, and the whole notion of causality as a uni-directional temporal sequence is ingrained in 'scientific' thinking. It forms the basis for such cornerstones of the scientific method as experimental design and the two-groups randomised control study, usually held up as the 'gold standard' for objective and scientific research (e.g. Ward *et al.*, 1992: ch.2). It is also the implicit bedrock for Popperian theories of scientific progress (see Popper 1959), insofar as a 'chain of causality' is the assumption underlying testing, confirming, or refuting hypotheses.

However, the fact that causality can be modelled following Blalock in terms of sophisticated regression equations is not 'proof' that causality exists in any physical sense. They do not model the world; they model our own thought processes. Thus, modelling the 'determinants' of delinquency in terms of statistically identified predictors does not fix delinquency as a consequence and the 'determinants' as the causes for all time. This is because the initial question, 'What are the causes of delinquency?' is itself an act of choice; the question itself reveals how we have *a priori* decided to segment the world. If we did the sums, we would find that amongst other things drug use was a predictor ('cause') of delinquency. But we could equally well ask, 'What does delinquency cause?' and use delinquency as the independent variable (rather than the dependent variable) in a similar set of regression equations. In such a case, delinquency is now the cause and not the consequence; and one of the things it 'caused' would be drug use.

Unfortunately, there are also further problems of a dialectical nature. It is argued here that not only is it the case that any element in a 'causal chain' is simultaneously a cause, an event and a consequence, but that even the temporal ordering of these into a series of recurring sub-cycles with no definable starting point (but always in the same order) is also illusory. Once conscious or volitional beings enter the equation situations arise in which backwards causality appears to take place. Indeed, where people are concerned it is arguable that the circumstance in which actions are caused by things which have not yet happened is in fact the *norm*. The snooker player's actions, for example, are a result both of what just happened (forward or conventional causality) but also, and crucially, what might happen next. The fact that a particular shot might *in the future* result in a snooker by the opponent causes the shot to be changed; the fact that it might be possible to pot the pink next determines where we shall try and locate the cue ball now. The game of chess is similarly dependent on backward chains of causality. The thing that characterises the expert player is precisely this ability to allow his/her actions to be 'caused by' states of affairs that have not yet arisen. By contrast, the poor player is he/she whose actions are merely consequences of (i.e. 'caused by') what has happened in the immediate past (pot this ball because it happens to have ended up near a pocket; take the pawn because it's possible to do so, etc.). Furthermore, the latter strategy regularly produces states of affairs that are *maladaptive with respect to future events* and is a good way to lose either game. Reacting to the past does not therefore mean that one is in any sense better equipped to face the future.

However, it can sensibly be argued that the expert player's actions are in fact caused by *internal representations of possible future events*. Such *internal representations* would in fact precede (and thus conventionally 'cause') the action; so the appearance of backward causality is, in this instance at least, probably illusory. We concede this point, whilst still entertaining the possibility that the *representation itself* might be explained in terms of backward causality insofar as it represents a future event that has not yet been experienced, at least not in precisely that exact form.

However, and more importantly, there are problems with the way in which the term 'internal representation' is commonly construed, especially in those cases where attempts are made to model 'internal representation' in ways which are isomorphic with researchers' assumptions about real world processes. Such 'mental models', (particularly the popular 'boxes-linked-by arrows' models, e.g. Fishbein and Ajzen 1975[10]) in our view simply model the external logic of the researchers' own ideas, rather than any internal process that actually exists in anyone's brain.

These reservations are discussed in detail in a later chapter, but it is sufficient to say at this point that we have greater confidence in models which have more fluid properties, particularly connectionist and neural-net conceptions (see Chapter nine). It should be noted however that in the absence of *any* sort of anticipatory mental activity, preventive action is logically impossible. In some form, people have to envisage things in order purposively to prevent them; whilst the actual word 'representation' may lead us up the garden path towards too literal an interpretation of what such an 'envisaging' process might be like, *something* has to happen in the brain prior to the event. We prefer a fluid conception of such a process, based (in phenomenological terms) on emerging concerns, dimly perceived possibilities, intuitions, hunches, and feelings derived as a result of the way in which prior experiences impact on, and change the properties of, neural substrates. This is slightly different from a view that sees the process as a monolithically logical/representational one in which fully formed systems of 'if ... then' gates, arrows and boxes, spring immediately to 'mind' in the form of fully articulated 'mental models'.[11] We concede the point therefore, that backward causality, taken literally, does not exist in a hard form in this instance since *the anticipatory imaginings of the accomplished safety manager* predate the events that might happen. But rather than being a set of literal internal 'representations' bound together with an inexorable systems logic, this capacity to imagine is (in the first instance at least) loose, fluid, dynamic, and evolves as a consequence of experience and the effects of that experience on connectionist type processes. This, in our view, would represent a key attribute/ability for anyone involved in preventive safety.

Justifying proactive safety

One might be forgiven for suggesting, somewhat cynically, that accountants have fewer problems in justifying massive expenditure in the wake of the last catastrophe than in liberating more moderate funds to prevent catastrophes from happening in the first place. The fact that the ability to imagine that ' something might happen' may be the only thing that precedes an accident *that was thereby prevented*, creates quite a problem when it comes to justifying safety in specific economic terms. Indeed, a system that actively and successfully avoids something does so because someone has identified a set of causes that are never allowed to cause anything and imagined an event that consequently never happens. Consequently there are *no hard data on specific things that have been prevented*. Statements such as 'We prevented two train crashes last Wednesday lunchtime, and a major explosion at 1.15 pm on Thursday' are thus impossible to 'prove' with data, even if they are (in some unproveable sense) true. Paradoxically, the effective system is thus harder to justify than less effective systems, because of the impossibility of distinguishing that which simply failed to materialise from that which was actively prevented. Less effective systems in unsafe settings are, ironically, thus probably the most likely to get funded because i) the evidence that they are necessary is there for all to see and ii) it is easier to show a reduction in accidents/incidents when there are a lot happening. By contrast, suppose a sweet shop runs an expensive incident reporting and root cause analysis system, and there are no serious injuries or fatalities. Is this because the safety system is so good, or because sweet shops are so intrinsically safe they just don't need expensive safety systems? And what if the same state of affairs obtains in a nuclear power plant?

On the other hand, safety systems which primarily respond to things that have gone wrong are easier to justify 'objectively', although they are by definition merely reactive and have to be judged ineffective from the point of view of prevention. Like the bad snooker player, they react to what has happened but fail to anticipate what might come next. 'Too much and too late' is the model from which safety systems need to escape. Learning from prior disasters is the least effective, economically by far the most costly, but (unfortunately) historically the most potent, model. They have one important advantage over proactive systems, however. They produce 'hard data' that impresses auditors and accountants.

We have briefly commented on the primarily reactive nature of safety systems which prevent things from happening *again*, and suggested a paradox whereby a deficient system under which things continue to go wrong can actually be easier to justify than a highly effective one. However, there is an equally worrying possibility which derives logically (reciprocally) from this argument, concerning those events which have *never* happened; this involves the development of what we have chosen to call a 'Precipice Culture'[12], because the process resembles deliberately approaching closer and closer to the lip of a dangerous precipice in order to discover at what point one falls off. The suggestion is that with the passage of time, the fear of a catastrophic incident *which has not yet happened* gradually declines; in a sense, familiarity with constant risk breeding contempt. Along with this decline in justifiable fear grows the suspicion that, since IT has never happened, some of the safety procedures around the potential event may in fact be unnecessary, and may thus be cut back. We suspect that an unspoken and possibly unconscious hypothesis-testing procedure may then commence, whereby the limits of safety are 'tested' by successively withdrawing barriers and safety procedures so that the imagined event becomes progressively more likely and in the end may actually happen. In a sense, this involves relying on the emergence of 'hard evidence' that you have got it wrong, at which point it is too late. This suggestion is impossible to 'prove' but we may note expedient actions taken by the managements of high-consequence industries involving reduction in manpower, double-tasking, re-skilling, changing of roster patterns, 'streamlining of procedures', expansion of service intervals and so forth which may insouciantly prove (eventually) to be part of such a process. Chernobyl may illustrate such a process of progressive barrier removal taking place within a very foreshortened time scale; but a similar process may take place over a period of months or years, and be much harder to spot since the rate of change is so much slower.

However, there is a way out of this dilemma once we stop treating systems of cause and effect as purely scientific concepts, realise that the sense of causality is phenomenological, and that the terms 'knowledge' and 'hard data' are not synonymous. The stock conference question, 'How effective is your system in reducing accidents?' reflects the preoccupation with the balance between empirical facts and safety measures, implies a reactive rather than a proactive orientation, and is misconceived. The assumption is that the *only* things which need preventing are things which have not been prevented previously. If the speaker is unable to point to a downward trend or a reduction, he/she is on thin ice even though a causal connection between a safety system and a reduction in incidents is in any case incredibly hard to demonstrate. As Groeneweg argues, the relationship between actual incidents and overall company safety can be extremely loosely coupled (see Groeneweg 1996: ch.3, for a detailed coverage of the problems associated with using recorded incident statistics as a general index of company safety).

What would be judged by many to be a quite unacceptable, possibly an ironic, response to the question 'Have accidents gone down since your system was introduced?' would be the answer, 'My safety system is successful because it has prevented certain mental images I have from actually turning into empirical evidence'. And yet, ultimately, this is *exactly* what proactive safety systems are for. They seek to reduce the probability of frightening mental images becoming observable facts. In other words, they represent the attempt to stop nightmares coming true. They depend on the human capacity to imagine and foresee consequences of actions undertaken, to calculate subjective costs and benefits in both economic and human terms, to make *moral* judgements about where the latter outweighs the former, and in some cases to take whatever steps are necessary to try to ensure that something *never* happens. This is a uniquely human and subjective process; something that people are good at, something which no computer can or will ever do, and is a key component of the safety process. It is also a component that is regularly underestimated by those who see the process entirely within a deterministic or 'scientific' framework. The fact that disasters often produce wonderful data is hardly an endorsement for having failed to stop them happening. Despite the claimed advantages of 'evidence-based' action, there are after all types of hard empirical evidence that we don't want and just don't need.

At the level of actual practice, this consists most of the time of nibbling away at the less spectacular individual links of imagined causal chains, so that each link removed can no longer serve as a connection in the imagination. Thereby *one* possible sequence of events is removed from the vast array of possible interactions that might conceivably combine to produce an accident. This is entirely consistent with Hume. *If a cause is 'an union in the imagination', then removing causes consists of removing imaginable links so that the notion of cause is no longer imaginable within that specific context.*

It also follows that identifying and removing links that could occur in a number of causal chains will, other things being equal, be more useful in reducing incidents than removing links which are highly specific to a single chain of events. The endeavour, however, is not *in the first instance* driven by empirical or observable facts, but by the imagination and intuition of the individual concerned in imagining causal chains. The key variable is therefore, 'How good is this safety expert at imagining things that are likely to happen?' The first step in good safety management thus has as much to do with inspired acts of imagination as with 'science' as normally conceived. No matter how empirically based, concrete and sophisticated our subsequent analysis, how complex our quantitative risk assessment, the first and most important step rests on the quality and intuition of the safety manager in deciding priorities. By contrast, safety management solely on the basis of data and empirical evidence is by definition always reactive and the response is frequently too much and too late.

One of the best ways to capitalise on the human capacity to perceive developing patterns and to draw anticipatory inferences is through the detailed use of minor event reports; that is the collection of data on incidents or situations which do not result in major consequences, and the addition of these to more traditional safety databases. Within that process, the main focus would be on people's *perceptions* of causes, events and possible consequences, and the best way to access these and add them to the armoury of the safety manager is through the principled and replicable analysis of what people actually say in their own words. That is, the analysis of natural, minimally cued, discourse.

This only makes sense, of course, if the causes of minor events are to be found amongst the causes of more major incidents, so that fixing minor event causes will prevent at least some of the more major incidents. The chapter on 'triangle' models in this text deals with that issue and confirms that this is the case to a useful extent. The role of the safety manager within this process requires that he/she has the vision and foresight to imagine which causes of minor events are most likely to form links in a causal chain which could result in a major catastrophe in the future, and which are less likely to form such links. In other words, to foresee the possible consequences of things that only had a minor impact on a particular occasion and to sort the wheat from the chaff. And that is a matter of subjective judgement as much as of science.

The prize for getting this correct – that is predicting that which can reasonably be predicted (and bearing in mind that not everything can be predicted in our uncertain and sometimes chaotic world) – is actual major incident avoidance. The approach contrasts with the more common approach of identifying dozens of unprioritised causes or contributory factors which tends to result from highly detailed investigation of the last disaster, whilst according low priority to minor incidents from which proactive lessons might be learned. Prevention thus requires imagination, in contrast to *Learning from Experience* (a perhaps unfortunate title, e.g. NHS Scotland 2002). Learning to avoid accidents *without* having to undergo the experience clearly has much to recommend it.

3 Predictive validity of near misses

Introduction

In Chapter two the issue was raised as to whether near miss analysis is valuable for the prevention of more serious accidents – this chapter aims to address this issue. Decision making about investing in safety improvements is usually based upon the relative importance of root causes in accidents and failures. However, such decisions can only be reached reliably by referring to statistics from large databases. As serious accidents themselves are generally (fortunately) few in number, they are of limited value as an aid to such decision-making processes, and therefore the use of near misses to increase data in databases is potentially one way to counteract this problem; but only if there are common causes for both serious and minor event/near misses.

Before discussing in detail the issue of whether or not the causes of major events are to be found amongst the causes of near misses and minor events, however, the reader should note the discussion of the notion of 'cause' given in Chapter two. The use of the terms 'cause' and 'causality' within the present chapter takes place within that construction; namely that a 'cause' is a psychological entity rather than a purely physical phenomenon.

The background to the common cause hypothesis

The use of near misses to aid the decision-making process is based upon the assumption that near misses and more serious accidents have the same underlying causal patterns. If the hypothesis holds good, then the causes of major events are to be found among the plethora of causes underlying more frequent minor events, and detailed scrutiny of minor events is warranted as a method of removing the causes of major incidents. This hypothesis leads immediately to the question, is it in fact the case that the causes of major events are to be found amongst the causes of minor events? A second question, concerning how one locates from amongst a plethora of minor-incident reports those low-level causes that are most likely to precipitate major events, is tackled later in this text. First, however, it is necessary to establish the validity or otherwise of the common cause hypothesis.

The assumption of the predictive validity of near misses was first posited by Heinrich (1931) and is generally known as the 'common cause hypothesis'. This common cause hypothesis is, however, not without controversy. Both Petersen (1978) and Saloniemi and Oksanen (1998) support an alternative causation hypothesis which suggests that the causes of near misses are different from the causes of accidents – in this case the collection and analysis of near misses becomes an end in itself and not a way to reduce accidents. In this chapter it is argued that Heinrich's common cause hypothesis has been misunderstood and therefore improperly tested by a number of authors. It is obvious from the literature that there has been a fundamental confusion between frequency (how often certain types of accident occur), severity (the consequences of the accident) and causation (the factors which combined to result in the accident). This has led to several authors inappropriately rejecting, or failing to support the validity of the hypothesis, as well as to studies which seek to support the hypothesis with data inadequate for that purpose. It is apparent that to date no adequate test of the common cause hypothesis has been performed. A study which attempts to rectify this omission and address some of these misunderstandings is described later in this chapter. However, first it is necessary to review in detail some of the work which has been carried out previously on this issue.

Heinrich first propounded the common cause hypothesis in his seminal book *Industrial Accident Prevention* published in 1931. This book used a number of arguments to demonstrate that accident prevention should be targeted at accidents that did not result in major injury, as well those that did. Heinrich attempted to show that reducing minor injuries and 'no injury' accidents would result in cost savings and increased productivity for industry. The rationale was as follows.

First, he demonstrated that due to the hidden costs of accidents, accident prevention made sound business sense. By reference to actual industrial accidents he showed that the incidental (hidden) costs of accidents was four times as great as the cost of compensation and medical bills. Heinrich further asserted that the hidden costs of accidents applied equally for major injuries, minor injuries and 'no injury' accidents (i.e. at a ratio of 4:1 of the actual costs). Thus prevention aimed at near misses as well as actual accidents could potentially lead to large cost savings. In 1966 Bird re-calculated the figures and came up with a ratio of 6:1. However, this serves only to emphasise the fact that accident prevention (if effective) could easily save employers much money.

Secondly, he discussed the interconnectivity of safety and productivity. He quoted figures from the American Engineering Council (1928) that a safe factory was eleven times more likely to be productive than an unsafe factory, thus highlighting the importance of safety management.

Thirdly, he examined insurance claims and suggested that for every accident resulting in an injury there were many other similar incidents that resulted in 'no injury'. He estimated that for every 330 potential accidents,

Figure 3.1 Heinrich's ratio triangle.

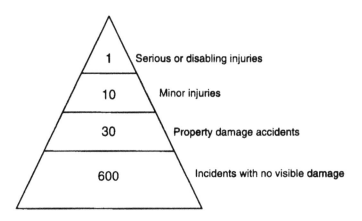

Figure 3.2 The Bird accident ratio study.

one will result in a disabling injury, 29 in minor injuries and 300 will lead to 'no injury'. Heinrich's ratio triangle is shown in Figure 3.1. Heinrich used his ratio triangle to suggest that a reduction in 'no injury' incidents would lead to a reduction in minor injury incidents and in major injury incidents. However, this would only hold if the ratios were stable and hence predictive. Since Heinrich calculated his ratio triangle, many other authors have replicated his study and arrived at different figures. Bird (1966) examined 1,753,498 accident reports from 297 companies. His results are shown in Figure 3.2. Swain (1974) found the same ratios as Bird, while Skiba (1985) provides a set of ratios which suggest that for every fatal accident there are an estimated 70,000 near accidents. However, while the ratio results are interesting in their own right, they are not related to causal patterns, they are merely descriptive of the inter-relationship between frequency and severity.

Fourthly, Heinrich postulated the common cause hypothesis based on his own analysis of 50,000 accident reports. He devised a taxonomy of causation

Table 3.1 Heinrich's early taxonomy of accident causes

Accident causes	
Supervisory	*Physical*
Faulty instruction: a) None b) Not enforced c) Incomplete d) Erroneous	Physical hazards: a) Ineffectively guarded b) Unguarded
Inability of employee: a) Inexperience b) Unskilled c) Ignorant d) Poor judgement	Poor housekeeping: a) Improperly piled/stored equipment b) Congestion
Poor discipline: a) Disobedience of rules b) Interference by others c) Fooling	Defective equipment: a) Miscellaneous materials/equipment b) Tools c) Machines
Lack of concentration: a) Attention distracted b) Inattention	Unsafe building condition: a) Fire protection b) Exits c) Floors d) Openings e) Miscellaneous
Unsafe practice: a) Chance taking b) Short cuts c) Haste	Improper working conditions: a) Ventilation b) Sanitation c) Light
Mentally unfit: a) Sluggish or fatigued b) Violent temper c) Excitability	Improper planning: a) Layout of operations b) Layout of machinery c) Unsafe processes
Physically unfit: a) Defective b) Fatigued c) Weak	Improper dress or apparel: a) No goggles, gloves, masks, etc. b) Unsuitable – long sleeves, high heels, defective, etc.

Source: Heinrich (1931)

which was divided into supervisory and physical causes. Beneath each of these superordinate categories he provided a number of classifications such as poor discipline, unsafe practice and physical hazards. Each of these classifications then had a number of detailed sub-categories from which to choose. We can see that Heinrich's taxonomy fails to identify 'supervisory' causes clearly, including as it does such things as 'faulty instruction' (which is a supervisory problem) along with 'inability of employee', 'mentally unfit' and 'lack of concentration', which are not under the control or management of the supervisor. Furthermore 'poor housekeeping' is included as a 'physical' cause despite the fact that this is clearly under the control of supervisors. However, although we may not fully agree with Heinrich's taxonomy it is nevertheless advanced for its time. Heinrich provided explanations and examples of the terminology he chose (for an overview of the taxonomy see Heinrich 1931: 46). A representation of this early taxonomy is shown in Table 3.1.

Using the taxonomy in Table 3.1 Heinrich applied his causal analysis to 50,000 accidents. We can see that the categories in Heinrich's taxonomy may be problematic. For example, an employee who is deemed to be physically or mentally unfit is categorised as a supervisory cause, while housekeeping is judged to be a physical cause. We discuss elsewhere in this text the problems that such a taxonomy can cause for reliability of coding. Nonetheless, Heinrich's taxonomy is historically important and is the forerunner of many taxonomies of accident causation today. The analysis of Heinrich's results shows remarkable similarities in causal patterns between major injury accidents, minor injury accidents and 'no injury' accidents. There are however, some unanswered questions regarding his data set. He does not provide information on how the causal analysis was performed. There are no reliability data and no information on who or how many individuals actually coded the data, although it is assumed that Heinrich himself performed the analysis. Further, Heinrich himself states that only one immediate cause was assigned for each accident – given our understanding of accident causation today, the data are somewhat limited. The data used appear to come only from the 'supervisory' causes, while only the mechanical hazards have been included from the 'physical' causes. The graph in Figure 3.3. shows Heinrich's results.

Despite these problems, the data provide compelling evidence for the similarity in causal patterns between major injuries, minor injuries and 'no injury' accidents. In the 1931 edition of his book Heinrich makes the following assertion:

> the predominant causes of no-injury accidents are, in average cases, identical with the predominant causes of major injuries – and incidentally of minor injuries as well.
>
> (Heinrich 1931: 90)

It seems clear that Heinrich has demonstrated that 'no injury' accidents (called near misses today) have some similarities to major and minor accidents, at least in terms of the immediate cause. It is rather surprising therefore, that in his later editions of *Industrial Accident Prevention* Heinrich no longer promotes this view, and in fact the common cause hypothesis is very much ignored in subsequent editions. Then in the final

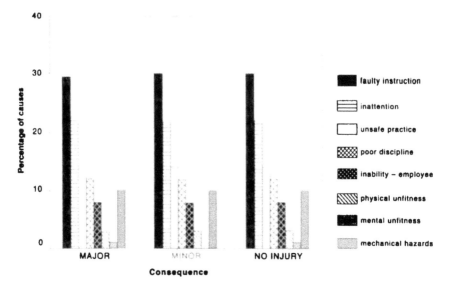

Figure 3.3 Heinrich's data on causes of major, minor and 'no injury' incidents.

edition of his book, as we shall see, Heinrich and his co-authors perform a very convincing (if somewhat misleading) U-turn on this very issue.

Arguments against the common cause hypothesis

Arguments against the common cause hypothesis have been raised by a number of authors (e.g. Petersen 1978, Heinrich *et al.* 1980, and Hale 2000). Some of these objections are examined in this section.

Heinrich *et al.* (1980) and Petersen (1978) raise the same set of objections. Their arguments are based on the observation that a reduction in frequency of accidents does *not* lead to a similar reduction in severity. They state that this is evidence against the common cause hypothesis. The following statements are made to substantiate their claims:

> We can readily see that the types of accidents that result in temporary total disabilities are different from the types of accident that result in permanent partial disabilities or fatalities ... handling materials accounts for 25% of all temporary total disabilities and 21% of all permanent partial injuries, but for only 6% of all permanent total injuries and fatalities. Electricity accounts for 13% of all permanent totals and fatalities but accounts for a negligible percentage of temporary totals and permanent partials. These percentages would not differ if the causes of frequency and severity were the same. They are not the same. There are different sets of circumstances surrounding severity.
>
> (Heinrich *et al.* 1980: 64–5; Petersen 1978: 19–20)

and

Statistics show that we have only been partially successful in reducing severity by attacking frequency. In the last 40 years National Safety Council figures show an 80 percent reduction in the frequency rate. During that period the same source shows only a 72 percent reduction in the severity rate, a 67 percent reduction in the fatal and permanent total rate and a 63 percent reduction in the permanent disability rate.

(Heinrich *et al.* 1980: 65; Petersen 1979: 23)

Hale (2000) also uses these figures and ratio related data to make a case against the common cause hypothesis. However, as we shall demonstrate in the next section, these arguments are misconceived and based on a confounded view of the common cause hypothesis.

A confounded interpretation

These arguments against the common cause hypothesis are based not on actual causes of accidents and incidents but upon their conforming to the original triangle model as proposed by Heinrich. The existence of a ratio relationship between major injury accidents, minor injury accidents and 'no injury' accidents says nothing about causation. Furthermore, the ratio relationship has been shown to be different for different industries and for different countries, and there has been a fundamental confusion between the relationship between severity and frequency and causal mechanisms. This appears to have been recognised in the final edition of *Industrial Accident Prevention* (1980) where Heinrich and his co-authors state:

It [the ratio] does not mean as we have too often interpreted it to mean, that the causes of frequency are the same as the causes of severe injuries. Our ratios and figures in this area have confused us.

The ratio relationship as first described by Heinrich was not intended to demonstrate causation, but to demonstrate that by taking action to reduce the frequency of accidents the severity of the accidents would also be reduced. This is independent of the common cause hypothesis. However, the reduction of severity in line with a reduction of frequency requires the ratio relationship to be stable and therefore predictive, at least within a given industry. Although many different authors have tested the ratio relationship there is no evidence that the ratio relationship is indeed stable over time. None of the studies performed (e.g. by Heinrich 1931, Bird 1966, Skiba 1985) have been re-tested at different periods of time, but have simply demonstrated the existence of a particular ratio at a given time. Such ratios may therefore merely be descriptive (i.e. a 'snap-shot' of a state of affairs), rather than demonstrating any causal relationships.

However, whether the ratio relationship described by Heinrich (1931) and others is stable and therefore predictive has no real bearing on the common cause hypothesis. The common cause hypothesis is concerned with causal patterns and is independent of any specific or claimed ratio relationship.

The empirical evidence evaluated

This confusion between the ratio relationship of accident frequency and severity with the common cause hypothesis can be demonstrated by a number of studies that claim to have tested the common cause hypothesis by using ratio data and either found the hypothesis wanting or lent their support to it. Such results and tests using inappropriate ratio data have then been used by other authors to substantiate their claims for or against the hypothesis. It is argued that a proper test of the common cause hypothesis has not yet been performed. This is despite Hale and Hale (1972) highlighting the fact that the hypothesis had not been empirically tested. The following sections highlight problems with ratio data that have been used to test the common cause hypothesis and with causal data. In each case a few examples of each are provided to illustrate the point (for a fuller review see Wright 2002).

Ratio data studies

This section illustrates the use of ratio data (i.e. data on frequency and severity of events) to test the common cause hypothesis. It is however, inappropriate to either refute or support the common cause hypothesis using data of this type – such data can only demonstrate whether the ratio relationship as first proposed by Heinrich has been replicated and do not have any implications for the common cause hypothesis.

Tinline and Wright (1993) reject the common cause hypothesis based on their research of two types of incident: LOCA (Loss of Containment Accident) and LTIs (Lost Time Injuries). They showed there was no correlation between the two types of incident either within or between companies. However, this analysis uses frequency and severity data and not causal data. Therefore this research cannot justifiably make claims about causal patterns or the common cause hypothesis, but only about the ratio relationship of the events. This move from frequency and severity to causal patterns appears to arise from the confounding of the causes of frequency and severity with the causes of incidents.

Similarly, a 1998 study by Saloniemi and Oksanen claimed to study the causal chains behind fatal and non-fatal accidents in Finland. Using only frequency data, the authors concluded that the common cause hypothesis was not supported and that fatal accidents are very different from non-fatal accidents. However, what their data actually show is that the ratio (iceberg) model is not validated for the Finnish construction industry. It would appear that the authors have confused causal pathways with the ratio rela-tionship produced by the iceberg model. As their data do not fit with Heinrich's original 1 – 29 – 300 model, the authors conclude that this is evi-dence of different causal patterns.

Both of these studies use frequency and severity data (i.e. how often seri-ous and non-serious injuries occur in a given industry) to dispute the common cause hypothesis. Yet neither of them analyse the accidents and incidents in terms of causes or causal patterns and then compare them. It is clear that these and other similar studies are not in fact testing the common cause hypothesis, but are in fact, testing the iceberg/ratio models. We shall turn now to studies that claim to use appropriate data (i.e. causal data) to test the common cause hypothesis.

Causal data studies

Shannon and Manning (1980) compared 2,428 accidents as reported by employees of the Ford Motor Company in England. Accidents were differen-tiated in terms of whether time had been lost from work or not. The accident causes were assigned by employees and a random sample was then validated by safety engineers. Manning's accident model (1974) was used to plot the data. This model is not a taxonomy of causal factors, but considers an acci-dent as a sequence of events, rather than as a single event, and is formed of a number of components such as 'an object', 'a movement' or 'an event'. It was concluded that lost time accidents have different causes from non-lost time accidents. However, despite these claims it is clear from reading the paper that accident causes have not been compared. The authors have compared the number of events that preceded the different types of incident and the objects which caused the injury, rather than the actual causes of the accidents and incidents. This paper sheds no light on the common cause hypothesis.

Unlike Shannon and Manning (1980) Lozada-Larsen and Laughery (1987) conclude that the common cause hypothesis is supported by their study of 7,131 employee accidents. They compared 6,435 minor injuries with 408 major injuries. The data appear to have been coded by the com-pany and there are no details of consensus or inter-rater reliability. Firstly, they examined the frequencies of various activities that were taking place prior to the accident occurring. These activities included such things as assembling/disassembling equipment, manually operating valves and han-dling materials. For both major and minor injuries the activities prior to the accident were strikingly similar in terms of percentages. However, whilst these data provide an interesting insight into the kinds of activities taking place prior to the accidents occurring, they do not shed light on causal mechanisms. For example, 'manually operating a valve' is not the cause of an accident – the cause is related to the way in which the valve was oper-ated. Was the valve turned the wrong way? Was the procedure followed? Was the wrong valve operated? Was protective equipment used? In this case the authors have failed to differentiate between the cause of an accident and the routine tasks performed by the employees.

This first test of the hypothesis was followed by comparing what the authors call 'accident events' with 'incident events'. These included such things as 'impact with an object', 'cut by', 'caught between'. While these data may be useful to industry in terms of the type of accidents and injuries that occur during normal, routine activities they are not valid for testing the common cause hypothesis. Heinrich (1931) is clear that for example 'impact with an object' is the cause of the injury sustained, but it is not the cause of the accident or event that caused the impact. Such data do not provide causal explanations but merely a description of the injury event.

Criteria for a proper test

As can be seen there is little transparent evidence of the common cause hypothesis having been properly tested at all – authors have instead tested the ratio model or have compared data that are neither causal nor ratio. A proper test would consist of gathering appropriate accidents for analysis; near misses, property damage incidents, minor injuries and major injuries. Following the collection of the data, the accidents should be analysed and causes assigned (not descriptions or precursor events or the object that caused the injury) according to an appropriate (i.e. methodologically valid) and reliable taxonomy. The causal patterns can then be compared for the various levels of 'no injury' accident, property damage accident, minor injury accident and major injury accident.

Testing the hypothesis

Such an empirical test of the common cause hypothesis was performed by Wright (2002). This section briefly describes the study and the results obtained. Data from the study all came from one large U.K. railway company. Accidents came from three sources: voluntary reports of near misses via the Confidential Incident Reporting and Analysis System (CIRAS) between 1996 and 1999; SPAD (Signal Passed at Danger) investigations; and Formal Inquiries of railway accidents. The severity levels chosen to perform the test of the common cause hypothesis were the following:

1   near misses (akin to 'no injury' incidents);
2   property damage accidents (this level of severity was later proposed by Bird in 1966),
3   fatalities/injuries grouped together.

It was not possible to distinguish between major and minor injuries from the collected data as the severity of injury was not recorded by the company, and therefore all injuries were combined. Table 3.2 shows the final data used to test the hypothesis.

Table 3.2 Data source and level of consequence

Consequence	Data source			Total
	Formal inquiry	SPAD investigation	CIRAS report	
Fatality/injury	17	0	0	17
Damage	18	7	0	25
Near miss	11	81	106	198
Total	46	88	106	240

Source: Wright (2002)

Technical and human factors causal codes were assigned to all incidents in the same way regardless of source in accordance with the University of Strathclyde CIRAS Human Factors model. This model is shown in Figure 3.4. The causal codes were assigned to the individual incidents on the basis of reading the reports thoroughly, constructing a causal tree of the event and then assigning the appropriate codes. Causal codes were only assigned when it was clear from the reading of the text that the particular code contributed to the propagation of the incident. The coding was performed by the author who devised the taxonomy in conjunction with the implementation of the CIRAS project. A reliability trial was also performed on the data. Index of concordance was calculated using the following formula:

$$\frac{\text{Number of agreements} \times 100}{\text{Number of agreements} + \text{number of disagreements}}$$

The number of agreements between raters was 29 and the number of disagreements was 8. This resulted in an index of concordance of 78.4 per cent (see Chapter eight for a discussion of the issue of reliability).

Codes were not assigned as either present or absent. Rather the number of times a code occurred was assigned (i.e. if two pieces of equipment failed and thereby both were causal factors in the development of the incident this would be coded as Technical Change × 2). This provides frequency data for each possible code.

The model in Figure 3.4 provides an overview of the logic of the CIRAS system. The individual codes are shown and described in Tables 3.3 to 3.5.

Proximal factors

Proximal causes of events are defined as causes associated with frontline operators (e.g. drivers and signallers). These causes are closest in time and often place to the occurrence of an event. These and additional proximal or 'sharp end' factors are shown in Table 3.3, with examples of the type of error made by the operator (as reported to CIRAS).

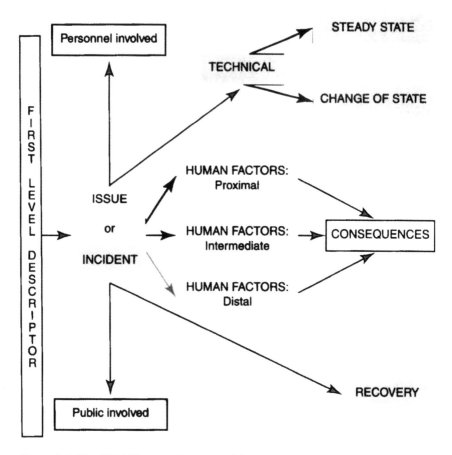

Figure 3.4 The CIRAS human factors model.

Intermediate factors

Intermediate causes of events are defined as occurring between proximal and distal causes. Staff involved are somewhat removed in time and place from the occurrence of an event (e.g. supervisors and maintenance workers). Intermediate causes include the following: maintenance (inadequate or not carried out), communication failures from managers to staff and training. These and others are shown, with examples in Table 3.4.

Distal factors

Distal factors are defined as organisational or systems related failures. Staff involved are likely to be removed in time and place from the event e.g. managers and designers. The distal causes of events are shown with examples in Table 3.5.

Table 3.3 Proximal human factors

Proximal code	Definition	Example
Attention	Lack of concentration leading to slips (physical) and lapses (mental/cognitive)	Failed to stop at station as singing in cab
Perception	Inability to see or hear specific features	Unable to see signal due to foliage
Knowledge	Lack of knowledge/inadequate or incorrect knowledge for task	Trainee unaware shunting procedure not authorised
Rule violation	Deliberate breach of rules or procedures	Not using electrical protection when necessary
Procedural error	Mistaken use of **wrong procedure in a given situation** (note the action is mistaken but not unintended i.e. the person meant to do the act at the time but didn't mean to get the rules **wrong**)	Opened doors on the wrong side of the train, but in correct manner
Slip or lapse	An unintended action	Driver slipped and fell when entering train
Communication between staff	Failure of communication between frontline staff	Driver failed to pass on relevant information to signaller
Fatigue	Tiredness/fatigue influencing behaviour	Train delayed as driver got out of train to walk along platform

Table 3.4 Intermediate human factors

Intermediate code	Definition	Example
Communication from staff to management	Failure of frontline staff to communicate with managers/ supervisors	Driver failed to report incident to supervisor
Communication from management to staff	Failure of supervisors to communicate with frontline staff	Manager did not provide feedback on reported incident
Maintenance	Inadequate or absent repairs/ maintenance	Monitors not maintained
Training	Insufficient for task, not provided	Training not provided on rarely performed task (isolating doors)

In addition to these human factors codes two Technical codes ('Steady State' and 'Change') were also included in the analysis. Steady State denotes a problem with equipment when working normally and Change denotes equipment that has failed.

The CIRAS human factors model is hierarchical. According to this model individual causal codes are subsumed under one of the four top-level categories: 'Technical', 'Proximal', 'Intermediate' and 'Distal'. The data for each level of severity for these top-level categories are shown in Figure 3.5.

As can be seen from Figure 3.5, the data are strikingly similar across severity levels. Proximal codes are the most frequently occurring human factors causes regardless of the severity of consequence. At the macro level, the coding hierarchy comprises a total of 21 individual causal codes. Technical comprises two codes, Proximal eight codes, Intermediate four codes and Distal seven codes. The fact that by far the greatest percentage of codes are Proximal, in all levels of severity, may be a reflection of the greater number of individual codes which comprise that category rather than a reflection of Proximal codes being more frequent in contributing to incidents. Intermediate causal codes account for approximately 10 per cent of the codes assigned to each severity level. Distal causal codes (akin to latent failures) account for roughly 20 per cent of causes in all severity levels. The largest difference between the severity levels occurs at the level of Technical codes, where there is a greater contribution of Technical codes to near miss incidents than to more serious incidents.

When the data were compared using Chi-square analyses for the difference between proportions as described by Fliess (1981) there were no significant differences between the three groups 'fatality/injury', 'property damage' and 'near misses' at the Technical, Proximal, Intermediate and Distal levels of the hierarchy.[1] In other words, there are no significant differences between the occurrence of different types of 'causes' at any of these levels, thus supporting the common cause hypothesis that there are no significant differences between the factors that 'cause' events at any of the three levels of severity.[2]

Table 3.5 Distal human factors

Distal code	Definition	Example
Top down communication	Failure of senior managers to communicate with staff	Failure to fully explain implications of restructuring
Procedures	Ambiguous, difficult to follow or absent	Procedure surrounding inoperative AWS open to interpretation
Design of equipment	Equipment design not fit for purpose	Lifting equipment for removal of train doors inadequately designed
Rostering	Shift scheduling, staffing	Short staffing requiring overtime, rest day working
Objectives	Management priorities	Performance before safety
Culture	Attitudes of work force, macho-style, general company ethos	Get the job done culture

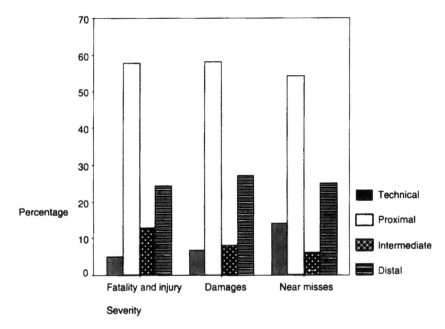

Figure 3.5 Technical, proximal, intermediate and distal causes by severity.

Similar analysis was performed on the total of 21 individual codes that the taxonomy contains. Results showed that of these 21 individual causal codes only three were significantly different across the three levels of severity compared. These causal codes were 'knowledge', 'training' and 'procedures'. In all cases there was a greater proportion of these three codes for the most severe accidents, indicating that serious events are more likely to involve failures of knowledge, training or procedures as causal factors, than are less serious events. There are a number of possible explanations for this finding. Firstly it is possible that in the case of more severe consequences these three codes occur more frequently and point to real differences in causation. If this is the case, then attempting to control and modify these particular causes should have an immediate and clear impact on reducing the severity of the incidents.

However, it is also possible that the special status of knowledge, training and procedures suggested by this analysis arises as an artefact of coding. In line with our definition of 'causes' as psychological rather than physical entities, it is entirely possible that these causal factors are ones that are most easily recognised by managers as causes that are easy to remedy after an incident occurs (they are fairly standard items that have traditionally dominated management thinking, and are therefore more prevalent when managers perform the investigation). Further investigation is needed to determine if these results are replicable over a period of time.

Whatever the true situation is regarding the specifics of the causal status of knowledge, training and procedures, overall these results provide support for the common cause hypothesis and point to the usefulness of collecting near miss data as a way to identify and prevent more serious incidents. The analysis we believe represents one of the few, possibly the only, direct test of the common cause hypothesis to date which makes use of actual assigned causal codes, as distinct from data on ratios and/or frequency statistics.

Collecting and analysing minor event reports is a useful thing to do

In Chapter two, a fundamental issue was raised concerning the extent to which the causes of minor events are to be found amongst causes of major events. If it is the case that common causes exist for both major and minor incidents, then tackling the causes of minor events may be expected to have an impact in terms of reducing major incidents. On the other hand, if the causes of major incidents are specific, then no amount of effort spent tackling the causes of minor events will have any impact on major incidents. On the basis of the present study, which involved a more detailed and painstaking approach to this question than perhaps has been previously attempted, the common cause hypothesis has been largely validated and it is concluded that the collection and analysis of near misses is a useful way better to understand the evolution of more serious accidents. The collection and analysis of minor event and near miss data can thus be expected to help an industry better understand how small failures can develop into more serious incidents, and therefore take action to prevent the latter. Therefore it is clear that the collection and analysis of small failures is an important component of company safety systems, with real implications for the avoidance of disasters. Furthermore as serious accidents rarely occur, the collection and analysis of minor event/near miss reports also serves the useful function of keeping safety in the minds of the workforce and of maintaining alertness (van der Schaaf 1991).

The way forward

This chapter has attempted to show that collecting and analysing small failures is worthwhile and can be used to help prevent more serious accidents and incidents. How then do safety managers get their hands on near miss reports? What is the best way to overcome the fear and embarrassment of reporting? The next chapter addresses the issue of how to collect near miss reports.

4 Confidential reporting as an approach to collecting near miss data

The nature of 'causes', and the way in which precursors to events achieve the psychological status of 'cause', has been discussed in Chapter two. The previous chapter then attempted to demonstrate that the collection and analysis of near miss or minor incident reports is a useful and valid way of determining the 'causes' of both minor and major events and enables preventive measures to be implemented. This chapter now addresses the problems of incident reporting systems, and suggests ways of maximising the benefits of such systems through confidential reporting.

A general coverage of the advantages associated with near miss and minor event reporting systems is given in the text *Near Miss Reporting as a Safety Tool* (van der Schaaf 1991) which provides a comprehensive discussion on setting up and maintaining reporting systems, along with case study examples. However, this chapter does not intend to cover the same ground but instead aims to outline the factors that should be taken into consideration when instituting one particular type of near miss reporting system, a confidential reporting system. The factors are highlighted by reference to experience with CIRAS (the Confidential Incident Reporting and Analysis System) that was first introduced for the U.K. railway industry in 1996.

Why confidential reporting?

Confidential and non-confidential reporting schemes share the same objectives, but confidentiality provides the reassurance that people need if they are to report many of the events or circumstances of most interest to risk management; particularly those concerning the reporter's own personal errors, slips and lapses, the details of which can often provide clues to systemic error-promoting conditions. Reporting often takes a great deal of courage, to overcome the fear of ridicule, embarrassment and retribution. This makes maintaining the confidence and trust of reporters essential. Harrison (1991) gives the following advantages of confidential reporting schemes:

1 improves the completeness of accident and incident reports;
2 helps overcome the barriers associated with near miss reporting;

3   increases the information to build up a human error database;
4   increases suggestions for improvements.

The advantages of confidential reporting over anonymous reporting

There has been much discussion in the literature about the issue of confidentiality versus anonymity in reporting systems. Like confidential schemes, anonymous reporting schemes aim to overcome the barriers associated with self-reporting of errors and accidents. However, the disadvantages of anonymous reporting outweigh the benefits gained. This is because anonymous reports cannot be followed up to obtain further details, nor can the source be verified. Therefore such a system is more likely to attract 'nonsense' reports or personally motivated reports designed to settle scores. Also, an anonymous system is reliant upon the initial report for all details – any missing information remains missing. This can have a deleterious effect on analysis and therefore upon subsequent attempts to address the causes of incidents. Conversely, a system that provides confidentiality can verify the source and perform interviews with reporters to elicit further details. There are, however, limitations to the confidential approach: reports cannot be verified by witnesses or by other staff who may have been involved in the incident, nor can technical evidence be requested from the company. Such limitations, however, are balanced by the willingness to report and the type of information provided by reporters.

The majority of third party reporting systems e.g. CHIRP (Confidential Human Factors Incident Reporting Programme, CHIRP 2002), ASRS (Aviation Safety Reporting System, ASRS 2002) and MARS (Marine Accident Reporting System, MARS 2002) all provide confidentiality rather than anonymity for the reasons stated above.

Management support

In order for any reporting scheme to be successful and generate sufficient reports it is necessary that it be given management support at the highest level of the company (Lucas 1991) and that management are seen to attend to and act on the reports received. Ives (1991) provides evidence of management 'killing off' a reporting system through the unnecessary re-organisation of a successfully operating system. It is therefore important that management understand the purposes of the reporting system and foster its use rather than creating barriers.

Management attitude to near miss reporting schemes

Lucas (1991) cites three factors that are directly under management control which are vital for the success of near miss reporting (not necessarily only for a third party scheme, but also for in-house near miss reporting initiatives).

These are anonymity, forgiveness and feedback. In the next section, we discuss these three systems properties, and give them qualified support.

Anonymity

Whether a near miss reporting scheme should be confidential or anonymous depends on the goals of the system, the depth of information required and the way in which information is reported. If the reporting system allows anonymous reports to be submitted, then information gained from the first contact is the sum and total of all information that can be gained – an anonymous system prevents re-contact with the reporter. Thus whilst agreeing in principle that the identity of reporters needs to be protected at all costs, we suggest a confidential system instead of total anonymity. In a sense, confidentiality demands a degree of trust and responsibility from both parties. By contrast total anonymity provides a carte blanche for the reporter which is open to abuse. In a confidential system, the reporter knows that the report is being taken seriously, as a follow-up interview will take place where further details may be required, and thus he/she is more inclined, in turn, to take the business in the serious way intended and to appreciate their active role in a valued process. Accordingly, CIRAS (which will be discussed later) is a confidential system which enables the reporter to be re-contacted and further information obtained. This ability to re-contact and obtain further discursive detail over and above the half-dozen lines usually allowed for 'in your own words' reports is extremely valuable.

Forgiveness

Once again, whilst agreeing with the general principle that blame-tolerance is an essential pre-condition if a reporting system of the type described is to work properly, the precise term 'forgiveness' does not quite capture the necessary dynamic. Forgiveness and disciplinary action are not mutually exclusive. It is, for example, possible to discipline someone and then, in true public school style, forgive them afterwards. On the other hand, it is possible to take one's cue from Tess of the d'Urbervilles, and take no action against a misdeed, but not forgive the perpetrator either. Finally, forgiveness, along with apology, might be very gratifying in interpersonal terms but achieves nothing with regard to putting things right. In place of any literal interpretation of forgiveness, therefore, we prefer a definition which simply stresses that management will not take punitive or disciplinary action against reporters no matter what the circumstances of the report. When the reporting system is confidential, this is taken as given. Much attention has been given to the punitive aspects of reporting near misses and accidents via a company or third party reporting system. Confidential systems are introduced in order to circumvent the disciplinary process in an attempt to gain information about what was happening 'at the coal face' rather than to find out who was breaking the rules or not performing adequately. Such systems are often no-blame, in so far as individual reporters are guaranteed not to be disciplined if they report something that they have been involved in only via the confidential system. However, if they are involved in an incident which is already known to management via their own internal reporting procedures and for which disciplinary action would be the normal outcome, the fact of subsequently reporting it to the confidential system does not thereby secure immunity from normal disciplinary procedures. In other words, whilst confidential reports are treated confidentially and there is no blame or discipline attached to such reports, should the incident come to light via other means, then the individual will be treated as they would normally, whether a confidential report is filed or not. In fact, however, if the system is correctly run this problem never arises, as management is never provided with the names of individuals making reports. There is thus no mechanism whereby reports through the confidential channel can be cross-referenced with reports or incidents that come to light through normal channels or investigations.

When ASRS was first introduced it provided immunity from prosecution for those submitting reports. This had the effect of generating reports on the same incident by a number of individuals present at the time of the incident – all thereby ensuring that they were immune from the disciplinary process. Immunity from discipline can be counter-productive to the aims of a confidential reporting system as it distorts the number and quality of reports received (duplicate reports are counted separately). This results in statistical evidence that is based not on the number of near misses actually occurring but on the number of individuals who witnessed or were involved in the same incident. Thus immunity from discipline should not be an integral part of any near miss reporting system.

Much discussion has taken place as to whether a reporting scheme should be no-blame, provide immunity or simply exist within an enlightened culture. Berman (1996) suggests that a no-blame culture is both difficult to achieve and potentially self-defeating, and proposes instead a culture of enlightened response, although he fails clearly to identify how this could be achieved. Reason (2001) also discusses the need for an enlightened or 'just' culture specifically in relation to the railway industry. He suggests that a no-blame culture is neither feasible nor desirable. He does, however, acknowledge the difficulty of reaching the state of a just culture, which depends upon understanding motivations as well as actions. In such a situation, managers have to decide which acts deserve punishment and which do not, and where to draw the line. This can in itself create what appears to be an unjust system. Furthermore, unless staff are able to discuss their motivations and actions openly, it is difficult for managers and those in a position to make judgements on the underlying motivations and required punishments to perform their function without prejudice or bias.

Marx (2001) also discusses the issue of a just culture in relation to the reporting of medical errors.

The desirability of a just culture is not disputed here, but the feasibility of moving directly from a blame culture to a just culture seems overly optimistic. Until it is clearly demonstrated that a just culture exists, and until employees trust their managers with information without the fear of recrimination, near miss reporting will remain an under-utilised resource. A number of authors (Adams and Hartwell 1977; Webb *et al.* 1989) have linked under-reporting of incidents and suppression of information to the apportionment of blame and disciplinary action i.e. to a blame culture. In the U.K. railway industry a combination of the blame culture and staff perceptions of the utility of making reports has resulted in under-reporting in the past. Clarke (1998) found that the pattern of intended under-reporting indicated a 'risk management' cultural approach (Lucas 1991) in the industry, which served to emphasise specific types of incident – especially Signals Passed at Danger (SPADs) and Wrong Side Failures (WSFs), which often tend to 'report themselves' in the sense of being highly visible, at the expense of other equally important incidents which could shed light on a broader range of potential problems. Clarke (1998) also found that only 3 per cent of drivers would report rule breaking by a colleague. Confidential reporting is the first step towards regaining the trust of staff and establishing a reporting culture. A 'just' culture is hopefully the next step.

Feedback

The third factor which Lucas (1991) identifies as vital for the success of near miss reporting is feedback. Reporting schemes may be readily accepted and embraced by staff – reports may be forthcoming in the start up phase. However, if staff are to continue to use the system and keep making reports then they have to see concrete results and receive feedback on what actions their reports have generated. This may be done through individual feedback or via a publication/news sheet which all potential reporters receive. This ensures that reports are not seen as entering a 'black hole' and disappearing without trace.

Incentives for reporting

A near miss reporting scheme may wish to increase the number of reports it receives by providing incentives for staff to submit reports. Such promotional campaigns should not be undertaken lightly, and the dangers of doing so should be taken into account. The rewarding of reports may lead to biases in the data which would not otherwise be present; trivial reports may be made in order to claim the rewards, or fallacious reports may be generated in order to receive rewards. Also, when the rewards cease, reporting rates may fall – thus confounding the reasons for the fall in reports.

A further problem can arise if negative reports (reports of things *not* going wrong) are encouraged. An example from the literature may help to highlight the dangers of rewarding employees for making reports concerning safety per se. The case, concerning smallpox, is taken from Makin and Sutherland (1991). The international health organisations undertook a concerted campaign to eradicate smallpox; the 'front line troops' for the campaign were health visitors. Each of these had a geographical area for which they were responsible. In order to motivate the health visitors a bonus scheme was introduced. Arguing that the final goal was eradication of smallpox, a scheme was devised whereby each visitor was rewarded according to the absence of smallpox in their area. However, although the visitors consistently earned good bonuses, smallpox remained endemic. When considered from the visitors' perspective the reasons for this apparently paradoxical situation become clear. If you are rewarded for lack of cases, there is an obvious incentive to turn a blind eye. Less mischievously, in such circumstances many individuals undergo a 'criterion shift' for reporting (in signal-detection terms) as a consequence of which cases which were previously seen as clearly positive become increasingly uncertain. Either of these processes leads to a mantra of the type, 'When in doubt don't report'. The system is also open to more mercenary and obvious types of abuse. Management finally realised the potential for both reporting bias and abuse, and the reward system was turned on its head. Instead of being rewarded for the absence of cases, visitors were now rewarded for finding cases. The results were dramatic: undiscovered cases now came to the attention of the authorities and could be treated. This case highlights the dangers of providing rewards for making reports, and particularly for negative reporting, and therefore such reward systems should not be part of an event reporting system.

Preparation and planning

Before a reporting system (confidential or otherwise) can be launched in a company there must be adequate planning and preparation. This includes providing a route via which reports can be submitted, (decisions include whether it is electronic, paper-based, form-driven), adequate staff and adequately trained staff should be provided to deal with the reports (this includes interviewing and investigation techniques, familiarity with the analysis process), selecting an appropriate analysis method, publicity to ensure staff are aware of the scheme, and the designing of a feedback channel.

Each of these issues is now discussed with reference to the CIRAS system, one of the largest confidential systems to emerge in recent years.

The CIRAS reporting system

CIRAS is the Confidential Incident Reporting and Analysis System currently being used in order to identify and deal with human factors problems on the railways in the U.K. CIRAS was initially a response to the contribution of human factors (including human error and latent failures) to incidents, situations and near misses on the railways, and first commenced operation in embryonic form in Scotland in 1996.

Why CIRAS was introduced

An earlier background report by Heybroek (1995) of Vosper Thornycroft, commissioned by ScotRail Railways Ltd, pointed out the role of human factors in the rail industry, and the importance of these has also been highlighted in other industries (e.g. the off-shore oil industry; the nuclear industry). Furthermore, existing official reporting procedures are often associated with disciplinary action, and this distorts both the nature and number of reports received. This is particularly true in the rail industry where, historically, relationships between workforce and management have sometimes been characterised by mutual mistrust and animosity, rather than co-operation. This results in a tendency for reports to become focused on technical failures and chance happenings (the reports tend to be strategic, defensive and external) with the human element being virtually absent. In some instances, it may even be the case that a near miss or incident with no obvious consequences will be deliberately concealed (i.e. the person concerned feels lucky to 'have got away with it this time') due to the perceived disciplinary implications, rather than being seen as something from which others could usefully learn (Frese and van Dyck 1996). The aim of the system is to collect reports from individuals (drivers, signallers and other safety critical employees) of near misses, incidents and error promoting conditions, which would not normally be reported through official channels, and to use this information to enhance existing safety management systems. CIRAS is not intended to replace existing reporting procedures, and is a complementary system which operates in parallel with existing reporting channels. The philosophy of the CIRAS system is that it is better to know what is happening at the 'sharp end' than to know who is doing it.

CIRAS planning

The planning of the CIRAS system began a year prior to the launch of the system. The design and operation of the system was based on best practice from other reporting systems and industries. At the time of the development of CIRAS, CHIRP (Confidential Human Factors Reporting Programme) had already been in operation for the aviation industry for more than a decade. Accordingly, the team visited CHIRP to discuss the implementation and data analysis performed for the civil aviation industry. Best practice in terms of feedback to staff via a newsletter was adopted following this visit. CHIRP do not accept anonymous reports (for reasons of clarification, obtaining further information and to prevent spurious reporting) and this was also deemed to be the best route for the CIRAS system to take. As the aviation industry was more advanced than the railway industry in terms of reporting systems, the research team from the University of Strathclyde also visited BASIS (British Airways Safety Information System). Unlike CHIRP or CIRAS this is an in-house reporting system, which has three tiers of reporting – one of them being a confidential system. A full description of the BASIS system is given in O'Leary (2001). Again valuable lessons were learned from the experience of BASIS. Finally, an administrator of EUFORCE (Captain Paul Wilson) kindly discussed his experiences and provided valuable information on the pitfalls to avoid and the necessary elements for any confidential system. A special committee was established to make decisions regarding the policies and processes which defined the scheme. The company concerned had already decided to implement a confidential system and to have an external third party (the University of Strathclyde) administer the system. The advantages of this meant that the company retained control of their data, but were not concerned with having to train analysts or interviewers. Further, the confidentiality was perceived as being absolute as no-one from the company had access to the reports or database derived from them.

It was decided to make the reporting system paper-based with the option of telephoning a report in to the CIRAS team as well. The original form contained a number of questions and tick box options, but it was soon changed to one question ('Tell us what happened and why you think it happened?') as reporters found it difficult to fill in the form. As all reporters were interviewed following their initial report or telephone call, the absence of a series of questions did not lead to the loss of any data.

A liaison committee was established to receive the information contained in the near misses, to provide feedback to staff and to determine the actions that would be taken on the basis of the reports. This committee included members of management, the trades unions, the Health and Safety Executive and the CIRAS analysts. Such a committee ensured that the system was well understood and that action was taken as appropriate. It further demonstrated to staff the seriousness with which their near miss reports were taken. This committee met on a three-monthly basis, although in cases of urgent reports committee members could be contacted between meetings.

CIRAS publicity

In order to ensure that potential reporters were aware of the system and the types of report that could be made, a series of briefing days was performed jointly by CIRAS personnel and the company concerned. Management demonstrated their commitment and support and the CIRAS personnel described the system, provided examples of the types of incident that could be reported and were useful to the system and then answered questions. These sessions lasted two to four hours depending on the number of staff attending. The staff attending these meetings were mainly supervisors, and they then briefed the staff that they were responsible for. In order that the majority of staff who could use the reporting system, knew who the CIRAS personnel were, a series of safety briefings was performed by CIRAS staff after the reporting scheme had been launched. This involved travelling to the various depots where staff were gathered on a fourteen-week basis for safety training, attending these meetings and giving a short presentation on the CIRAS philosophy and how the system worked, to ensure the message was disseminated to the grass-roots level. In order to allow staff to speak freely, supervisors were not present during these meetings. In addition leaflets describing the system were produced and distributed to staff.

In order to ensure that all personnel had access to the CIRAS telephone number, special credit-card sized cards were designed with the CIRAS telephone number. All potential reporters were provided with three forms and pre-paid envelopes as part of their 'starter pack' and racks holding forms were made available in mess rooms (this ensured that a reporter did not have to ask his immediate supervisor for a form). Further forms were also provided with the quarterly *CIRAS Journal*.

CIRAS feedback

We have already discussed the importance of feedback in maintaining the impetus of a reporting scheme. CIRAS provides feedback on three levels. The first is at the level of all potential reporters. A three-monthly publication (*CIRAS Journal*) is produced which lists the reports and provides the response of the company concerned. This allows all potential reporters to read about the types of report received and about the actions taken by their company. Staff are also encouraged to write in and comment on the company responses.

The second level of feedback is at the individual level. This is not performed as a matter of course, as it is time consuming and due to the nature of the confidentiality, reporters details are only kept for a specified limited time period to allow for the follow-up interview to take place, before being destroyed. However, individual reporters are provided with an update on the progress of their reported concern if they call personally and ask for an update, and provided with an identification code assigned to their report at the start of the process.

The third level of feedback is to management as regards the number, nature and type of reports received. A report to management was (at the time of the original system) produced on a three-monthly basis and contained data analysis, graphical representation of causal factors, descriptions of the incidents received and recommendations for actions. As an aid to management a list of proposed actions and the status of those actions was also compiled. These reports were also reviewed by the steering committee who set up and oversaw the project.

CIRAS confidentiality

CIRAS is confidential and 'blame free', and therefore staff can report not only technical failures, but also operator or human errors without fear of recrimination and discipline. CIRAS does not provide immunity from discipline but has opted to be no-blame (that is no blame, discipline or punishment will arise from the submission or contents of a CIRAS report – however, submission of a CIRAS report does not provide no-blame status should the incident already be under investigation or later detected) due to the existing, often confrontational, culture within the railway industry.

The CIRAS system, by ensuring confidentiality, seeks to rectify this imbalance, promote a reporting culture and hence producing causal human factors data that otherwise go undetected and unrecorded. CIRAS is also timely since it complements the privatisation of the rail industry in the U.K., opening the door to new, more open management systems and changes in safety culture.

CIRAS operation

In the simplest terms, safety critical staff voluntarily report safety concerns, unsafe actions and practices direct to CIRAS personnel. Reports are made via a standard reporting form (which asks for a name, phone number and address, and information about the incident in the respondent's own words) or by telephoning directly to a CIRAS analyst. Reports are followed up by a telephone or face-to-face interview where more information is sought including demographic data on time, place, date, length of shift, etc. Interviews are not performed on either employer premises or time. Interviews are tape-recorded with permission and fully transcribed to facilitate the analysis and feedback of information to the individual companies. Figure 4.1 outlines the basics of the CIRAS process (see Davies *et al.* 2000; Wright *et al.* 2000).

Expansion of the CIRAS system

Following the success of the CIRAS system for the companies enrolled from September 1996, and the interest generated within the industry for such a system, it was mandated in June 2000 for the whole of the U.K. Railway Group. This, in effect, made it compulsory for all companies who were members of the Railway Group to join the system. The mandate was

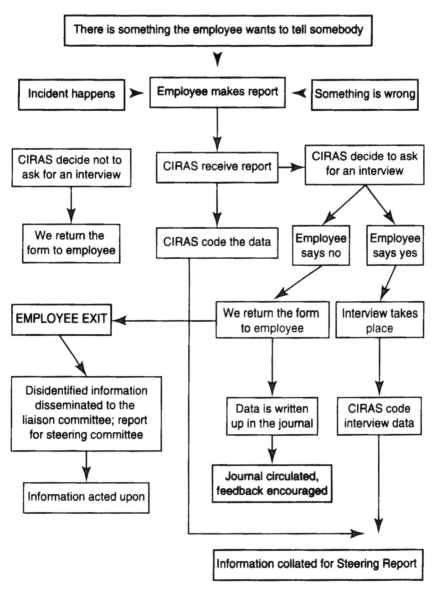

Figure 4.1 The CIRAS process.

applied by the Railtrack Safety and Standards Directorate who were then responsible for safety, Group Standards and the Rule Book. In order to move forward from a locally-based system to a nationally-based system able to handle reports from over 70,000 staff, as opposed to 5,000 involved at that time, the system was logically divided into a number of parts. The expanded system thus consisted of three data collection bases (Regions) and one analysis centre (Core). The Regions or data collection bases would receive reports and perform a cursory analysis, feed the information to liaison groups consisting of industry managers, trades unions and CIRAS staff and publish the responses and actions in a quarterly journal. Each of the three receiving bases would publish a separate journal to retain a local flavour which was found to be preferred by the staff reading the journals. The Core facility was to be responsible for the in-depth analysis and coding of all of the data collected nationally and for writing six-monthly industry reports. The various companies were divided amongst the three Regional facilities thus providing full coverage of the U.K. mainland. The Core was responsible for writing Standards to ensure that the CIRAS system provided a united front and a seamless entity, despite being run by various organisations. Each Regional facility works to the CIRAS process outlined in Figure 4.1 and all adhere to the same CIRAS standards and procedures to ensure confidentiality.

Conclusions

From a scientific perspective it is difficult confidently to assert a causal relationship between a confidential reporting system and increased safety due to the complex nature of intervening variables. However, there is no doubt that the information gained from confidential reports has impacted, and continues to impact, on specific changes with regards to safety in a number of industries. Consequently the correct perspective on confidential reporting systems is to see them as an important contribution to safety, culture and the better understanding of human errors that leads, in conjunction with a number of other factors, to progressive risk reduction. Furthermore the confidential channel produces information of a type which does not often come to light through official reporting channels and thus makes a unique contribution to understanding incidents and reducing risk.

What are the general prerequisites for a successful confidential third party reporting system? In this we cannot do better than to quote Charles Billings, who founded ASRS. He writes: 'Strong, widely held consensus that more and better information is needed is essential for the development of successful incident reporting' and continues 'The second requirement is for a respected body, one independent of the influences of other stakeholders, to conduct the collection and analysis of data.'

5 Numbers and words in safety management

Introduction

There are various references in this book to the use of textual accounts in safety management. We have argued that these accounts are acceptable as evidence, provided they are dealt with rigorously and systematically (see Chapters one and two). This chapter will extend this argument by showing that quantitative data have often been preferred to qualitative data on questionable grounds. The conclusion we have reached is that safety managers and researcher/practitioners should use *all* relevant and reliable data in pursuit of safety management goals.

Triangulation

The term 'triangulation' was initially borrowed from the realm of quantitative psychological methodology within a theoretical framework of psychological testing (Campbell and Fiske 1959). 'Triangulating' quantitative and qualitative data traditionally involves viewing these as distinct, and 'adding them together' to produce a 'complete' picture. This is called complementary triangulation and, in our experience, this is the method most often adopted by safety managers.

Complementary triangulation

Complementary triangulation arises from the view that interview results, for example, are 'just things people tell us', whereas 'official' accident statistics, for example, are 'objectively true'. This leads to the view that these data sets cannot be used to *test* each other, as the methods cannot be said to measure the same phenomena. The complementary approach typically involves using qualitative data to add 'breadth or depth to our analysis' (Fielding and Fielding 1986: 33). So we have the 'real' numbers, and 'illustration' in terms of what people tell us, which can be used to add to our general understanding.

Advocates of complementary triangulation criticise the idea that results produced by qualitative and quantitative methods (for example, interviews with staff, accident frequency data) can be *integrated* (i.e. used for *mutual validation*, e.g. Fielding and Fielding 1986; Flick 1992, 1998). Fielding and Fielding argue that integration relies on a false assumption of 'a common epistemic framework among data sources' (31). This established position has been criticised by those who advocate *validatory* triangulation.

Validatory triangulation

In simple terms, validatory triangulation involves using the degree of convergence between different data sources as an indicator of the validity of results. Denzin (1978) thus argues that findings from use of different methodologies can be regarded as more valid than a hypothesis tested only with the help of a single method. 'Triangulation', as another author puts it, 'reduces the risk of systematic distortions inherent in the use of only one method' (Maxwell 1998: 93). Many authors have outlined problems with complementary triangulation (in which the quantitative–qualitative dichotomy is maintained: see, for example, Robson 1993; Breakwell *et al.* 1995) which can be avoided with the validatory approach.

It is clear that in complementary triangulation, the relationship between quantitative and qualitative data is commonly that of *subordination*, or what Brannen (1992) calls 'pre-eminence of the quantitative over the qualitative' (24). The typical view is that 'qualitative data may [...] be useful in *supplementing* and *illustrating* the quantitative data obtained from an experiment or survey (Robson 1993: 371 *emphasis added*). Saludadez and Garcia (2001) argue that the dichotomous relationship between qualitative and quantitative research is maintained in order to establish a differential in status by turning the quantitative method into something it is not (i.e. objective and value-free, see also Laurie and Sullivan 1991: 126). (Hammersley [1992] argues that 'the distinction between qualitative and quantitative is of limited use and indeed, carries some danger' [39].)

The main danger is that the researcher gathers 'objective' numerical data and then *picks and chooses* discursive data to match. For example, suppose 'incidents' have risen from n = X to n = X + Y in a given period. If we simply accept this as *fact*, it is relatively easy to gather qualitative data that adds to this picture (the complementary approach). For example, operators might simply be told prevalence has increased and asked to explain why this *is the case* (no doubt they would outline some plausible risk factors [i.e. attributions for the increase] in this case). Note that the interview data are not tested against the numerical data. The acceptance of the numbers as 'true' means the words are gathered simply to back up this assumption.

To illustrate why this is problematic let us turn this scenario back to front. Suppose we first ask operators about incident prevalence, and they *say* they have increased from n = X to n = X + Y in a given period. Would we then simply go looking for numerical data to see *why* this is the case? Do many safety managers imagine that this is the approach they would take? Is it not more likely that we would be looking to *validate* the words by examination of the numbers, i.e. to test *what the operators have told us* by looking at the incident numbers.[1]

But in the first case we did not feel able to *test the numbers* in light of what the operators said. Of course this is because of the view outlined above (e.g. Fielding and Fielding 1986), whereby numerical data (in this case incident frequency data) are seen as epistemologically distinct from qualitative data (in this case verbal estimates from drivers). The latter, it seems to follow, are *more subjective* and must only be used to complement the former. This position seems worthy of investigation.

The epistemology of incident frequency data

To start at the beginning, 'incident' is a word. Moreover, it is a *categorical classification* to certain events: a bit like 'communication failure' or 'resource problem', only less specific. The word 'incident', however, is often treated as objective fact rather than subjectively applied 'code'. In order to show how this is inappropriate, a definition of 'incident' is required. There will of course be strictly codified definitions of many types of incident (plant events, near misses, minor events, abnormal events, 'code reds', health and safety events, dangerous occurrences, etc.) which are specific to different industries. However, the generic definition chosen should cover most cases.

Let us state that an 'incident' is *a deviation from accepted levels of operation*. Figure 5.1 shows this graphically.

Figure 5.1 shows an abstract picture of an incident. Note that the actual shape of the picture will vary, but the 'extent' of the incident can always be defined as the area above the grey shading (acceptable operations) and below the actual level of operations (the thick black line).[2]

Incidents are defined by detection and retrieval points. The 'detection lag' will vary but will be between the initial deviation from acceptable levels (ID) and the detection point (DET). The 'retrieval lag' (between IR and FR) will also vary. The example above has a 'stabilisation' period that may not always apply, for example a technological 'fix' may mean that the thick black line comes quickly back down once the detection point is reached. The incident ends when operations become acceptable again.

The important point is that the level of the 'grey area' (i.e. what is acceptable) is *arbitrary and socially defined*. 'Incident' remains a subjective category based on the level beyond which *consensus* states operations are unacceptable. What constitutes an incident is always what we agree to be one. If the incident is the area between actual and acceptable operations and the latter is contextually variable, then incident frequency data are no more or less objective than any other categorical data. In the example above, we can *redefine* the event as a 'non-incident' simply by raising the acceptability threshold, and so, whilst in practical terms there may be

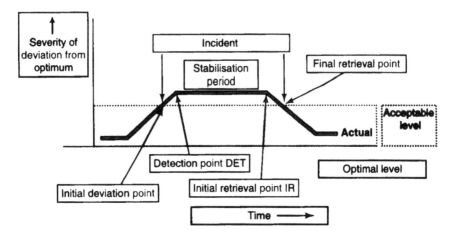

Figure 5.1 Anatomy of an incident.

'actual' values on the Y axis, the categorisation of 'incident' is a subjectively applied code.

For example, the Reporting of Injuries, Disease and Dangerous Occurrences Regulations (RIDDOR 1995), require reporting of injuries which result in the injured person being incapacitated from work for a period in excess of three days. The 'incident' classification relies on a *subjective* decision as to when this becomes reportable. It is not hard to imagine employees returning on the fourth morning (i.e. just prior to the initial deviation point) swathed in bandages whilst the company statistician asks 'what incident'?!! If we move the threshold, we alter the frequency.

It is of course true that, once established, the written 'rules' for classifying RIDDOR incidents are 'objective' if consistently applied. However, we would strongly argue that this is analogous to the rigorous approach recommended for the analysis of textual data in Chapter six on hermeneutics, where coders apply categories based on agreed written rules of interpretation. So an acceptance that incident frequency data are based simply on reliable categorical assignation means these can be seen as epistemologically identical to, for example, reliably coded pieces of text.

The deviation, detection and retrieval points which define an incident (see Figure 5.1) often appear 'objective' because the Y axis will equate to a *reliable* data source, for example temperature, fluid ounces, time, yards, spectra, etc. These data are difficult to argue with, and it may be accepted that these are particularly reliable and can be treated as 'objectively true' for pragmatic reasons. However, whilst this is true of the numbers per se, the other part of the definition comes from the acceptable level that is socially agreed upon (i.e. the *relationship* between absolute levels and set levels). So whilst we can generate abstract, reliable numerical data based on temperatures, times, etc., it can be argued that in safety management such 'absolute' values are never enough. We always want to know (i.e. detect) whether things are *too* hot, *too* wide, not strong *enough*, *too* dark, etc. (i.e. how many deviation points we have had). This is the principle behind 'control charts'. In meaningful data analysis, we always want to know how standards are being maintained. When we read, interpret and report on numbers, we introduce a subjective element.

Note that this does not lead to a *relativist* position where, for example, 'temperature' or 'weight' are illusory or entirely socially constructed. Because we can *agree* on them (i.e. they can be measured *reliably*), then actual levels can be used to differentiate between events. But once we begin to talk of 'high' temperatures or 'heavy' loads or volume being 'beyond control limits', then we can confer no special 'objective' status on our data. It is an axiom of safety management that we examine data primarily to check whether it is within set levels. Thus a certain subjectivity usually colours our view.

The implications of removing the 'special' status often conferred upon incident frequency data (and numerical data relative to standard) is that we can combine analysis of textual and incident-based data in the pursuit of safety goals. The following example shows how insight can be gained from cross-reference (i.e. validatory triangulation) between discourse and other data sources. In the case study described, discourse was used to test a hypothesis generated by numerical data, then numerical data were gathered to test *an alternative hypothesis generated from the first comparison.* This ongoing process, it is argued, makes the best use of available data, and allows for a scientific account of the system that accepts the role staff play in constructing organisational aspects.

Case study: Validatory triangulation in a safety management context

During annual data analysis, a major train operating company (TOC) found that a particular type of set, (traction unit A), seemed to be associated with signals passed at danger (SPADs). Over a five-year period, it was established that traction unit A was more likely to be involved in SPADs than any other unit (even when figures were corrected for traction miles).[3]

As argued above, the SPAD frequency data are partly subjective, based on consensus as to what constitutes a classification of 'SPAD'. However, it was accepted in this case that the numbers generated were reliable. Naturally enough, the differential between different traction units led to a perception that there may be a particular problem with set A. However, we did not view this as 'proven' by the frequency data. This was viewed as a *hypothesis* generated from the numerical data, rather than an established fact.

In order to test this hypothesis, focused discussion groups (e.g. Millward 1995) with train drivers were conducted, in order to cross-reference their opinions on traction unit A with the high SPAD frequency data for this unit.

Essentially, rigorous analysis of drivers' accounts did not confirm the original hypothesis. Traction unit A was seen as reliable. In fact, drivers preferred it to other sets. (Note that, if the interpretation of numerical data had been accepted as *fact*, it would have been relatively easy to gather data to 'illustrate' the original hypothesis. Drivers might simply have been told there was a problem with traction unit A and asked to explain this. Prompted in this way, no doubt they would have given us some plausible attributions as to why this might be the case.)

The cross-referencing approach taken in this case, however, highlighted a lack of consensus between the perceptions from analysis of incident data and the perceptions of the drivers. This lack of consensus led to generation of a *new hypothesis* whereby there might be unrecognised bias in the incident data (i.e. other, as yet unidentified variables that account for the numerical association between traction unit A and SPADs). This called for a further analysis of incident data, this time to test the hypothesis generated from *the comparison of the first two data sources*; this shows how a picture can systematically be built up.

Briefly, further analysis of quantitative event data produced findings which 'agreed' with the new hypothesis. Variables associated with the use of traction unit A (repetitive routes, station stops, etc.) were identified which it was felt might be the source of the differential in SPAD data (for a further discussion of how these factors relate to SPADs see Chapter ten, on arousal). In everyday language, the initial finding that SPADS ran at a higher level for a particular kind of set led to the natural, but basically misconceived, question, 'So what's wrong with these sets?' In the event, the qualitative data actually *disconfirmed* the hypothesis behind this initial question. The problem arose because of the way in which the sets were being used, and examination of *quantitative* data on routes and schedules *confirmed the qualitative data*. What was initially conceived of as a purely technical problem concerning sets and braking systems became transformed into a multi-level human factors issue concerning the effects of repetitive routing on driver arousal (i.e. nothing to do with the sets themselves). The company are currently testing the new hypothesis, derived now from i) numbers, ii) words and iii) more numbers, by examining train despatch procedures at stations on certain routes, and other aspects of use of these sets.

This short description of validatory triangulation in safety management shows the circular nature of the process. An increasingly reliable picture emerges as data sources are seen to be consensual or contradictory. Each data source (if *reliable*) is viewed consistently, and differences between sources are not 'ironed out' by assuming one source to be 'truer' than the other, but are incorporated into a new 'picture'.

Dealing with discourse

Finally, it is usually the case that, in this context, natural discourse is 'more time consuming and demanding to collect, analyse and make a case with' (Antaki 1988b: 12) than 'event' data. Qualitative data have been described as an 'attractive nuisance' (Miles 1979). The 'nuisance' refers to the legal doctrine that if you leave an attractive object, such as an unlocked car, where children can get access to it you may be liable for injuries they sustain. In Miles' analogy, qualitative data, left unattended by prescribed method, can 'injure' the unwary researcher who comes across it and is beguiled by its charms. Those who wish to work within a scientific framework must therefore take great care to deal with these data systematically. Meaningful analysis of qualitative data relies on reliable categorisation. This involves a process whereby coders must be trained to make decisions that can be tested for consensus. Harvey *et al.* (1988: 41) recognise that this process may not be straightforward, stating that: 'it is usually a laborious task to train coders to make useful discriminations and thereby to secure reliable judgments'.

It was established above that arbitrary categories often underly seemingly 'objective' frequencies which can be used for statistical analysis. One way to avoid the effort required to deal with discourse, of course, is to use rating scales or questionnaires to turn discourse into numbers. However, once more, it is problematic to view numbers generated from these as somehow distinct from and 'more objective' than discursive data (see, for example, Davies 1996).

In criticising the view that numerical data are always 'objective' we should not be blind to their attractions. We are aware that safety managers and researchers may often be under pressure to meet analysis and reporting deadlines and budgetary requirements. For example, Antaki (1988: 12) points out that what questionnaire users 'lose on the swings of validity, they gain on the roundabouts of convenience'. Convenience cannot be seen as irrelevant, and may sometimes take precedence for practical reasons. However, notwithstanding this advantage of numerical data, no system of any kind can be justified on the basis that it is merely *convenient*. It has to be shown to be effective and to work. A system which is highly convenient but achieves nothing is, obviously, of no use to anyone.

We prefer to argue that, where possible, all data sources should be examined to see whether they can shed light on the issues in question, and that both qualitative and quantitative data have equal status in that regard. Thus numerical data (e.g. incident frequencies and survey data) should be employed where they are appropriate and valid, not because it is decided that per se they are preferable (in terms of objectivity) to discursive data. Conversely, qualitative data have a reflexive and equal role to play in any such process.

Summary

This chapter has described how *complementary* triangulation (two different analyses adding up to give a complete picture) has often been preferred to *validatory* triangulation (two analyses of the same phenomena being used to test each other). It has been shown that this preference relies on an epistemological distinction between quantitative and qualitative data. Specifically, an 'objectivity' is conferred upon numeric data (e.g. questionnaire ratings, incident frequency data) and denied qualitative data (e.g. interview transcripts). This position has been criticised, and a safety management project is described which, it is hoped, illustrates the benefits of validatory triangulation.

In conclusion, safety managers are urged to forget the misconceived distinction between 'good' quantitative data and 'dubious' qualitative accounts, and to distinguish instead between reliable data, gathered and analysed by scientific method, and unreliable data, gathered by dubious method and analysed in an unprincipled way; and to use the former, both qualitative and quantitative, in pursuit of safety management goals.

6 Hermeneutics and accident reports

Background

Understandably enough, safety managers tend to take a fairly pragmatic view of their job. In accident investigation, experts study the accident and work out what happened: action is taken, and hopefully similar accidents are then avoided. Interviews can be carried out with key staff and information about matters such as 'safety culture' can be obtained. Accident reports come in, are filed, coded, and the data analysed. All these are part of the task of reducing quantifiable risks in the workplace, improving safety and, therefore, in the long run, helping the organisation save money.

But is it really that simple? One of the main purposes of this chapter is to show that it is not desirable to study 'risk', 'accidents' and 'human error' in the same way as we study valves and microprocessors. Instead, we have to accept that we always talk about these subjects *with (and therefore through) the medium of language.* At first this might seem a statement of the obvious, but acknowledgement of this fact has implications which are frequently overlooked in most discussions of safety management methodology. However, once it *has* been acknowledged then far more comprehensive analyses of data become possible. These should, of course, help to improve safety in the organisation and perhaps even make life easier for the person who has to analyse the data.

It is the intention here to focus attention on the added value to be gained from a rigorous and principled analysis of safety reports. However, this approach has implications for almost all aspects of safety. We are human beings, and because we are human beings we use language; and language, far from being an irrelevance that helps obscure our 'objective' view of the situation, is a tool without which we could not do anything at all.

Reporting systems

Increasingly, companies and organisations collect textual, qualitative (i.e. 'in your own words') reports of safety incidents. These are usually stored, sometimes as hard copy, and sometimes in a database. What to do with these reports is a common problem in safety/risk management. They contain information that might be useful in terms of accident prevention, but there is uncertainty about how are they to be analysed, and what *sort* of information might be obtained from them.

In a sense it is quite clear why incident/accident reports are so problematic, and why there is a preference for quantitative, numeric, 'hard' data, as opposed to qualitative or textual data. Seeing information in numeric form seems to imply to many people that the data are 'objective' and 'scientific', reminiscent of the sort of data produced in physics and chemistry. Qualitative data, on the other hand seem to raise all sorts of problems. The data are hard to analyse; moreover, there seem to be numerous competing methodologies for interpretation, some of which are mutually incompatible. Issues of reliability and validity are complex and are still debated, and the kind of data which are produced at the end of the process remain stubbornly linguistic, and open to objections that they are merely 'anecdotal'.

One of the themes of this book is that risk and safety management is a field in its own right and must develop its own methodologies. That is, it should not simply or solely aspire to the quantitative techniques of physics and engineering, especially in areas where these approaches are inappropriate. Dealing with natural text, what people say and write, is one of these areas. This does not mean that we are recommending moving over to the other extreme and suggesting that *only* qualitative techniques should be used. On the contrary, only by making the most of both approaches will real progress be made, but it is important to realise that both approaches have positive and negative aspects, that they do different things and that it is quite possible to extract the best elements from both approaches despite the different philosophical bases.

Positive and negative aspects

To begin with, a qualitative 'lump' of data, be it an interview, a written report, a transcript or recording of a focus group, is, in a sense, empirically more defensible than quantitative (for example questionnaire) data. It is a transcription of the actual words used by the subject; not someone else's on-the-spot individual impression about what they think it means, condensed to a sentence or a tick in a box for reasons that are personal and unknown.

However, it is the advantages of qualitative data that also lead to its disadvantages, in that it is precisely the fact that qualitative data are so unstructured that makes them so hard to analyse. The *quantitative* approach, on the other hand, has the overwhelming advantage of producing data which are amenable to statistical analysis: and statistical tests which permit 'trending' are particularly prized in a safety context. Moreover, it is often simpler in terms of presentation to demonstrate data in a quantitative form (for example, charts or graphs) than the 'messier' format of qualitative text (for further discussion of this see Robson 1993: 401–2).

We would argue strongly that the best approach for human factors/safety management in the context of error reports is one in which *both* qualitative and quantitative methodologies are used. Specifically, we would argue that the best and most accurate method of *gathering* data is qualitative (i.e. gathering the actual words of the subject), and the best method of *analysis* (in terms of presentation and 'trending') is, on the whole, quantitative. However, this is easier said than done, because in order to create practical methodologies of use to safety managers, we have to question some commonly held assumptions about the way human beings behave and use language. Before we can do that we have to challenge some assumptions about *texts*.

When a report of any sort comes in, be it a safety culture report, an accident investigation, a first person description of an accident, a transcription of an interview or whatever, it is undeniably a text first and foremost. It may be a text about any of the things listed above, and we may form a *hypothesis* as to who produced it, why, and 'what they were trying to say', but it is *a fact* that we hold a text in our hands. Part of the problem is that to many people this statement may seem at best banal, and at worst a statement of the blindingly obvious. But pressing further, we may discover that it seems obvious as a result of commonly held, but highly questionable ideas about language.

An example of such a view uses the so-called 'conduit' metaphor. Without having any empirical evidence to prove it we would still guess that this view of language is implicitly held by most safety managers and occupational psychologists, whether they know it or not. The conduit metaphor posits meaning in terms of discrete units, similar to physical objects. That is, I have ideas and concepts in the same sense that I have a heart or lungs: they are like objective physical objects, held 'internally'. When I 'tell you something' I have an 'internal process' which puts ideas into language. Language then functions as a pipeline or conduit, carrying ideas (unaltered and unedited) from my head into yours, whereupon it becomes 'placed' in your brain, where you can have access to it by another 'internal' process. This view is, as it were, the 'linguistic' aspect of 'cognitivism': the view of human cognition as being the algorithmic manipulation of 'internal' symbolic objects (see Chapter nine for a discussion of cognitivism).

This metaphor is so deeply embedded in Western views of the mind that many people have difficulty in seeing that it *is* just a metaphor (Lakoff and Johnson 1999). However, there is much experimental evidence that shows that it is not an accurate description of how people actually use language. Instead, it is better to think of language as being 'functional discourse': a kind of behaviour that we produce in certain situations to achieve certain goals. Language is not a pipeline or conduit, it is a tool. Its aim is not the straightforward transfer of the contents of one brain to another, but *to get things done* (for example, the question 'Have you got a match?', whilst semantically appearing to be a request for information transfer from one brain to another, is actually a request for a match). It can have many functions depending on the social context. Certainly it can be used to communicate, but it can also be used to persuade, to provide a 'take' on a situation which suits whoever is talking, as a shield behind which the person speaking can hide, to *mis*communicate, and so on (Winograd and Flores 1986).

If that is the case, then the status of texts becomes problematic. We can't delude ourselves any more that we can simply read a text and gain access to the thoughts of the person who produced it in the same manner by which we access information from a computer. Instead, a text becomes a repository of social meanings, produced in a certain context and read by us in a certain context for specific purposes. Moreover, there is no particular reason to think that the interpretation of a text carried out by an 'expert' is *necessarily* going to be any better than anybody else's. Instead, s/he will be reading the text in a different context for his or her *own* purposes, the same as everybody else. Reading becomes a complex activity, open to many biases and influences.

This view may seem controversial but it is actually only common sense. We all know that we read things in the light of our previous beliefs, opinions and experiences. We don't simply process them, behaving as if everything we read is 'true' or an accurate representation of the situation. We know perfectly well that people provide explanations for their own actions that may be 'less than the whole truth' and that accident reports and investigations are carried out by human beings with their own opinions and agendas. Certainly we have anecdotal evidence from certain industries that safety managers are well aware of these facts, and do not simply treat accident reports (especially those written by their own staff) as being unproblematic descriptions of objective reality. As we stated earlier, texts are primarily *texts*, not conduits through which thoughts are transmitted.

If we accept that this is the case, however, how are they to be interpreted? How are we to 'know' what is the correct interpretation, in terms of gaining the information we need to help improve safety?

The key problem is one of verification. If texts are not unproblematic carriers of information, but instead have to be interpreted, how is this to be done? In theory one could interpret the text and then go and ask the person who wrote the report (or the person who was interviewed, or both) whether one's interpretation was correct. But if language is really functional discourse (as argued in Chapter seven) then all we get in response to that question is more functional discourse of the same type. There cannot be two epistemologically distinct types of language, 'true' language and 'functional' language. And even if there are, we cannot make an arbitrary switch between language as 'truth' and language as 'function' just whenever it suits our purposes to do so.

Hermeneutics

Hermeneutics began as the study of religious texts, specifically the 'correct' reading of works such as the Bible or the Torah.[1] This was complicated by the use within those texts of parables and allegory, with the result that working out the 'correct' interpretation of these for the laity was one of the major functions of the priesthood. This of course begged the question as to how one would *know* that one had arrived at the correct interpretation, divine guidance being claimed as one of the guarantees of validity in that regard. Safety managers, however, are not usually members of the priesthood, and can hardly lay claim to divine guidance in the interpretation of safety reports, and thus have need for something more earthly to aid their endeavours. Nonetheless, hermeneutics raises two fundamental problems for both priests and safety managers alike: first, how to analyse texts in a way that looks beyond merely superficial features in order to understand and collect deeper meanings, and secondly, the problem of verifying the analyses so produced.

Hermeneutics remained a branch of theology until the nineteenth century, when the philosopher Wilhelm Dilthey (1833–1911) attempted to extend the subject beyond the religious domain and illustrate its importance for dealing with texts of many different kinds. Dilthey, following the theologian Friedrich Schleiermacher (1768–1834), posited hermeneutics as a way of defending the social or human sciences from what he saw as the encroachment of the methodologies of the natural sciences. It should be noted that despite the fact that he was a philosopher not a theologian, Dilthey's basic concerns remained the same as those stated above: to answer the questions, 'To what extent are readings an interpretative act?' and 'To what extent are they objectively "true"?' Twentieth-century hermeneutics oscillated between the idea that an interpretation was an *individual's, one-off* act of empathetic interpretation (relativist), and a more 'positivist' approach which emphasised the extent to which readings were social and structured, and, therefore, verifiable (Mallery *et al.* 1987).

It was the German philosopher Martin Heidegger who made the next major innovation in twentieth century hermeneutical thought. His major work *Being and Time* (*Sein und Zeit*, 1926) is extremely complex, but there are two major themes it discusses which are of interest here. The first of these is the extent to which interpretations presuppose other interpretations. As a grown adult one does not approach a text 'in a void'. Our readings of the text in front of us will always be coloured by our previous readings of other texts, in terms of the presuppositions, prejudices and assumptions we make. But these other texts were in turn interpreted by the light of other texts and so on. As Heidegger puts it: 'Interpretation is never a presuppositionless apprehending of something.' (Heidegger 1962: 191–2).

Secondly there is the paradoxical concept of the hermeneutic circle. When we read something, we 'break it down' into its component parts (words, paragraphs, and so on). But these parts only make sense in terms of the whole of the text. There is an endless dialectic, according to Heidegger, in terms of how we read, between reductionism (breaking the text down) and holism (reading it as a whole). This has two implications: one positive, and one negative. First, there is a paradox: we can never understand a text without breaking it down, and yet we must look at it as a whole before we understand the individual parts. For Heidegger, this is a 'chicken or egg' situation: each step logically presupposes the other. Interpretative reading is therefore a paradoxical action in that, logically, we can never begin to read, as we must always have begun the other step before the one on which we are now engaging. In order to escape this paradox, we must make a 'leap' of subjective (non-rational) intuition 'into' the text. To posit a merely rationalistic (non-subjective) approach to reading is paradoxical and incoherent (Heidegger 1962).

However, the correlate of this is that once the reader is 'inside' the circle, understanding can proceed via a dialectical process, and, therefore, progress. The text is broken down and then built up again: and each time this is done the 'meaning' of the text becomes clearer.

It is clear that the effect of Heidegger's thought was to relativise hermeneutics. If my reading of a text is coloured by presuppositions it is clear that it will be different from yours (with your different presuppositions), perhaps radically so. Moreover, according to Heidegger there is no way to bridge this gap. Reading is not a rational or scientific process, but instead begins with an intuitive 'leap of faith', and the form of my 'leap' will clearly be different from yours. The only progress I make in understanding is my own personal progress in the hermeneutic circle, but this is essentially a private, not public process.

The hermeneutic tradition since Heidegger has generally been split between those who have followed Heidegger in his fundamentally relativist viewpoint, and those who have instead stressed the extent that hermeneutics is a practical method which can produce 'objective' readings. To illustrate this debate we will briefly describe the thinking of Hans-Georg Gadamer (1900–2002) (who followed in the Heideggerean tradition) and Paul Ricoeur (1913–) who has attempted to create a hermeneutic *methodology* for deciding upon the 'correct' interpretation of texts.

The theories of Gadamer

In his classic work *Truth and Method* (published in 1960) Gadamer attempted to reinvigorate the hermeneutic tradition by re-emphasising its subjectivist elements (Gadamer 1981). So in this work he criticises the correspondence theory of truth (the idea that a sentence is true or not depending on whether or not it corresponds to 'the facts' about 'external reality' [Rescher 1973]). Gadamer's objection to this theory is that it does not take into account the force of the hermeneutic circle: for Gadamer *there is nothing except language*, the hearing and reading of which only presupposes more language. The assumption of some mysterious 'reality' outside language is a redundant and unprovable hypothesis to Gadamer (Palmer 1969).

Gadamer's approach may seem radical but it has a number of advantages for the qualitative analyst. First, it does not necessitate endless attempts to match discourse to 'reality', an infinitely complex task (for example, if every event report had to be followed up by another report to check its accuracy, since this report would also be a text, it would then have to be followed by another investigation and so on in a situation of infinite regress). Second, it does not support the concept of the 'expert', whose readings are 'better' or 'truer' than other people's (in other words, someone who can tell better than other people to what extent the text 'corresponds' to 'the facts').[2] There can be no expert readers for Gadamer, only various readings which have greater or lesser validity for the reader. The main disadvantage of his theories, of course, is that the spectre of relativism remains unexorcised: without the correspondence theory perhaps texts can be interpreted to mean anything one wishes.

It is Paul Ricoeur who has attempted to take up this challenge and create a hermeneutic methodology which acknowledges the strengths of Gadamer's arguments but which does not result in subjectivist relativism.

The theories of Paul Ricoeur

Ricoeur's method (which he has termed the hermeneutic arc) has two stages: first a move from the 'subjective' to the 'objective' and second from the 'objective' to the 'subjective', thus, according to Ricoeur, preserving the benefits of both approaches (Ricoeur 1981).

The first stage of Ricoeur's approach is to form a hypothesis as to the meaning of a text on the basis of subjective intuition. This hypothesis must then lead the reader to classify the text into a hierarchy of elements. The reader then moves from the part to the whole (from the individual classified elements to the whole text) and back again, in an attempt to 'get an idea of what the text is about' (on the surface). Despite the fact that Ricoeur admits in the final analysis that texts have potentially many meanings, and that various classifications may be valid, he insists that a social process analogous to that carried out by a court of law (and, not, therefore, analogous to the method of the natural sciences), can test these hypothetical distinctions. This is, therefore, a social process. He agrees with Karl Popper that genuinely objective statements must be 'falsifiable' but he does not accept Popper's belief in the 'correspondence' theory: according to Ricoeur a reading of a text is falsified if it is *internally* incoherent or the social 'legal process' rules it implausible – in other words, he agrees with most contemporary philosophers in the 'coherence' or 'consensus' theory of truth (Thagard *et al.* 2002; Rescher 1973).[3] In other words, does it describe something that people agree about, and if so, does it serve as a reasonable basis for action?

In the second stage of the hermeneutic arc, Ricoeur analyses the meaning structure of a text, which lies underneath its surface structure. Texts are seen to be composed of *meaning units* which constitute the whole text, in the same way as paragraphs are constituted of words and sentences. These are formal characteristics which can be studied only in relationship to other aspects of the text (Mallery *et al.* 1987).

So for example, in terms of a common narrative or story the most obvious example of 'meaning units' might be to break a text down into the 'beginning, middle and end' of the story. Or (again if the story or narrative was simple) you might break it into before and after an event. So in terms of a rather well known story you could break a fairy tale down into 'the section *before* grandma was eaten by the wolf' and 'the section *after* grandma was eaten by the wolf and little red riding hood arrived at the cottage'. Or the story could be broken down in any number of other ways. The key point, from our point of view, is that these 'meaning units' be practical (they have some bearing on the purpose of the interpretation), and that everyone can agree on them (that is, that after a process of discussion, everyone will be able to break down similar stories into similar discrete elements).

Of course the example above dealt specifically with narrative, but there is no reason that meaning units of this sort need to be temporal (beginning, end, before, after, etc.). They could be very different sort of divisions. For example the story could be broken down into 'parts of the story set indoors' and 'parts set outdoors' or 'sections where wolves were predominant' and 'sections where human beings predominate'. It really doesn't matter for our purposes, as long as the salient parts of the text can be so classified, the sections of text can be reliably assigned to the categories (see Chapter eight), and the categories have some sort of practical explanatory purpose. Then, again, the reader moves between these 'deep structures' and the whole text and back again, in an attempt to discern the underlying meaning of the text. This is the second and final part of the 'arc', and Ricoeur believes that via this method readers can achieve agreement on what individual texts 'mean' in a certain context.

The key point of Ricoeur's process, therefore, is to see the extent to which readers, each with their own differing perspectives (this view is termed 'perspectivism' and it is clearly derived from Gadamer), can achieve agreement over the meaning of a text. We have then added a final stage (described below), a *consensus trial* to assess the degree to which this agreement has been reached.

Ricoeur therefore believes he has solved the various problems of hermeneutics. Interpretation has been formalised, and (with the addition of a consensus trial) can be validated with the same degree of 'objectivity' as any scientific theory by seeing to what extent consensus has been achieved. However, his method does not presuppose that texts are unproblematic 'transmissions' of 'internal states' from other people (the 'cognitivist' hypothesis), and so does not lead to the paradoxes of 'expert readings' as argued by Gadamer. Instead it breaks texts down, and looks for the deeper, internal structures of meaning which lie inside them. In other words he views texts as social artefacts. Their 'truth' is the degree to which we can achieve consensus on their meaning. Moreover, in terms of the various oscillations between micro structures and macro structures he believes he has incorporated Heidegger's 'hermeneutic circle' in such a way that it facilitates analysis, not rendering it impossible.

So far problems and methods of interpretation have been discussed. What has not been mentioned is the *purpose* of interpretation.

The purpose of reading

It is clear that in the social world we actively seek out information. However, it is equally true that we wish to *use* this information. Texts are not just more or less meaningful, they are also more or less useful. It is important to raise this point because the question of 'objectivity' is difficult to answer unless one knows the social context within which it is being posed. 'What is the best reading of this text' really presupposes that we know 'for what purpose' and 'in what context'. Consideration of how we react to real world texts, for example, 'phone books, training manuals and so forth, reveals a practical attitude: that texts are used to accomplish specific goals. Even reading a book because it is 'interesting' (i.e. arousing) is a goal, and reading for arousal is no less reading for a purpose than any other kind of reading (see Chapter ten). This point has been raised by Gadamer, who points out that in medieval hermeneutic tradition the third aspect of reading (after understanding and interpretation) was *application*, and that a modern hermeneutics should return to this tradition and realise that the meaning of a text will depend on the concrete situation in which it is read, and the use to which it will be put (Gadamer 1981: 275). Therefore, in this view, there is no contradiction in positing a practical hermeneutics for a specific purpose: safety management for example.

However, this begs a final question: why use hermeneutics at all? The answer is that hermeneutics was created to deal with the specific problems of textual interpretation discussed at the beginning of this chapter; moreover, it is a unified and coherent tradition which attempts to deal with these issues. Therefore it differs from other interpretative strategies whereby the initial philosophical problems are not even stated or faced, for example Grounded Theory where 'the originators of grounded theory, Glaser and Strauss ... have, in their various works, in effect combined strands of pragmatism, positivism, phenomenology, and hermeneutics, without making any attempt to explain coherently how they tie together' (Rennie 1999: 6). Hermeneutics is above all a unified coherent approach to these problems: it deals with the problems raised at the beginning of this chapter by simply studying texts as texts, not as conduits of information from 'inside a person's brain' to someone else's.

Ricoeur's hermeneutics, therefore, is anti-cognitivist, and sees the truth of interpretation as the degree to which we can achieve a pragmatic social consensus on that 'truth', rather in the way the legal system works. It therefore fits in well with the 'systems' approach we are proposing as the new paradigm for safety management.

An organisational model of human factors

As discussed earlier, an effective analysis system should ideally be able to take advantage of both qualitative and quantitative approaches to data gathering. The question is: how does one take advantage of the usefulness of qualitative data whilst producing data which can be analysed using statistical techniques? This is an important point, in that hermeneutic approaches have been discussed in the safety management field before (Taylor 1987). However, apart from implicitly proposing a purely qualitative methodology (as opposed to the qualitative/quantitative approach proposed here), theoretical discussions such as Taylor's avoid the problem of reliability: how is one to justify one's own hermeneutic approach?

The CIRAS project

The specific approach here was developed towards the end of the 'roll-out' phase of the CIRAS project (see Chapter four of this text), between 1999 and 2001, as a way of obtaining maximum value from the incident reports that were coming into the system at that time. It is a system for obtaining confidential data about safety issues and incidents in the U.K. railway system. A number of studies have been carried out which show that confidential systems (that, is, systems where reports of incidents/issues are disidentified; where the aim is to find out what happened rather than who did it) can produce data which might be missed through more conventional channels (Lucas 1991).

It should be repeated that CIRAS was a confidential system; reports were disidentified, and so the core had no way of discovering the identities of reporters, or contacting them in any way.

So CIRAS is an example of a situation in which texts are read in a highly specific social context with a highly specific end in mind, namely to produce data concerning safety, especially human factors issues, in the U.K. rail industry. Moreover, CIRAS was *not* an accident investigation project. It was often difficult to 'verify' reports due to the confidential nature of the system. Therefore, even if it was desirable to do so, a system based on the 'correspondence' theory of truth frequently could not be used, as it was impossible to ascertain what the 'facts' actually were. Hence the emphasis on social consensus/coherence in the current project.

The following pages describe the attempt to develop a practical and reliable hermeneutic approach to the analysis of a year's worth of CIRAS data, to see whether such an approach would yield new forms of analysis and add value to traditional qualitative techniques. It should be noted that in the following paragraphs the data presented in Figures 6.1, 6.2, 6.4 and 6.5 are fictitious. They do, however, illustrate the kinds of analyses which can be performed on data from the system.

The initial process

The key problem we began with in terms of developing a hermeneutic methodology was first of all to create the 'rules' which the readers would have to 'follow' in an attempt to create genuine 'objective' (used in the sense that Ricoeur uses it: agreed upon) readings. There were two specific problems here. The first problem was to create social rules that stayed true to the hermeneutic circle, that is continually to create movement from micro-units (the individual words and sentences of the text) to the macro-unit (the whole text, taken in its entirety). The other problem was to ensure that readers would progress from surface features to deep features.

But, most importantly of all, this methodology had to be developed such that it could be agreed upon, and used by all the coders. In other words we follow Ricoeur, in stating that the key aspect of claiming an objectivity in terms of a text's meaning is 'the fixation of the meaning' (Ricoeur 1977: 328), but this methodology was to be worked out during discussions with all the coders, such that a unified methodology (which could be written down in the form of a 'rule book') could be created. How was this to be done?

First, we created a series of stages of interpretation (easily done with the NVivo computer software package) in which readers would move from the 'micro-elements' to the whole text. Secondly, we attempted to ensure that readers moved from 'surface features' to 'deeper structures' in the text.

Therefore, to begin with, it was decided to create a series of categories of 'surface' 'grammatical' features of the texts. These were simply of the form 'who' 'what' 'where' 'how-why'. In other words, sentences were simply to be read as sentences, with 'how-why' functioning as the verb . So a sample sentence such as 'I drove the set past the red light' would be categorised with 'I' in the 'who' section, 'the set' in the 'what' section, 'red light' in the 'what' section, and 'drove' and 'past' in the 'how-why' section. These are simply 'common sense' grammatical features at the surface level, and were created solely to create a 'close reading' of the 'surface' 'micro-units' of text. This could be done as the first stage of the process in the NVivo software package: with the elements of texts assigned 'coding stripes' (Figure 6.1 provides a fictitious example).

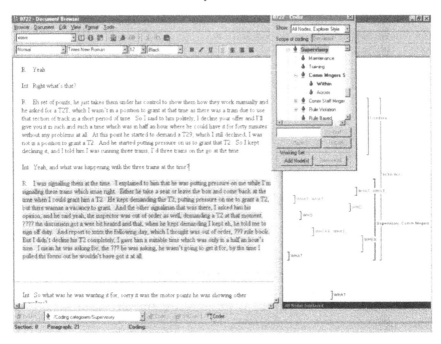

Figure 6.1 A screenshot of the hermeneutic process.

The next stage of the methodology was to read the text through again as a whole. Therefore in order to follow the 'rules' readers had to progress from breaking the text down, to building it up again – a continual dialectic between the 'smaller units' and the 'text as a whole': the hermeneutic circle.

Next the text was broken down into a 'deeper' level: this time relating to a general reading of the report. This broader structure is the 'frontline', 'supervisory' and 'managerial' distinction. Again, a 'group' judgement was made that the industry could reasonably be broken down into three layers, frontline, supervisory and managerial. Frontline issues were concerned with 'workfloor' issues at the 'man-machine' or 'man-task' interface; supervisory concerned with supervisors, and (self-evidently), managerial concerned with managers. It should be noted that this was a construct specific to this present project. For example in other industries with different data it might be possible to create a four-tier model, with an added distinction between lower ('middle') management and senior management. The categories were not set in stone, but were reflexive categories functioning as social 'rules' which were decided upon by debate, both within the CIRAS team and with the firms/companies involved.

At this point the entire text was again reviewed and frontline, supervisory and managerial coding stripes applied; typically these would identify larger sections of text, e.g. paragraphs. Therefore each relevant section of text was assigned a position at the frontline, supervisory or managerial levels. (NB: Each element of text might have been of any size from sentence to paragraph depending on its internal structure and degree of coherence. It should be noted that not all the text could be so broken down, and that only sections of text which could be so treated were 'carried on' to the next stage of the process.)

The next stage in the interpretation process takes the textual elements from the database and repositions them in a graphical format. Graphics sheets were created. These resembled 'flow charts' as used in engineering; however, they described social relations, (as posited by readers) between elements of texts, not physical causation between mechanical objects. Text fragments were taken and entered onto these charts such that a graphical representation of the situation could be created. Finally, coders assigned a code to the fragments of text which, from the graphical representations, were seen as being the most salient in terms of the purpose of the project.

Within the system there are 105 discrete codes, which describe a wide variety of human factors, demographic, and other safety related data.[4] These codes are contained within a 'coding sheet' which shows the codes in an easy to use hierarchy. The codes entered on the sheet (which may be paper or in an electronic format) were entered into an SPSS database. Where the 'coding sheet' is in electronic format, it was possible to input data automatically to an excel-type database, and from there to a statistical package for analysis. If not, data were inputted by hand to the statistical package and then transferred to excel-type software when graphical representations of the data were required.

The human factors codes were inferred from a three by three matrix which cross referenced the frontline, supervisory and managerial distinction by a distinction of activities being concerned with job/task, communication or procedural. This was the key point of the process, in that it was the codes created from these divisions that were to be analysed. These three distinctions were created in order to be inclusive but mutually exclusive. In other words, in the same way as all staff members could be categorised as either frontline, supervisory or managerial, it was argued that all 'error' activities could be categorised as being either job/task, communication or procedural. These were created from a logical *a priori* division of activities within the rail company which presupposed that staff enter into a pre-existing management/organisational structure. This structure had to possess codified rules of behaviour (rules and *procedures*). The activities which had to be performed within the structure were, logically, of two kinds: either 'physical' 'corporeal' *tasks*, or else activities concerned with *communication* (defined here simply as matters pertaining to discourse). Therefore, an organisation had to have *procedures* (*organisational*), which defined how staff carried out *tasks* (*job*) and *communications*.

It was this 'matrix' that functioned as a 'deep structure' to provide the final 'reading' of the text.

The most important point to understand is that these categories were created by discussion between all the coders, in terms of creating a set of categories that were mutually exclusive but all-inclusive; moreover, they were based on initial readings of a certain number of texts. By mutually exclusive we mean that definitions were socially created (through debate) such that each definition excluded any other. So part of the definition of 'procedural' errors was '*not* job/task' and so on. This methodology was derived from linguistics, where, as Ricoeur argues: 'it is always possible to abstract systems from processes, and to relate these systems … to units which are merely defined by their opposition to other units of the same system.' (Ricoeur 1977: 334).

The definition as being all inclusive was again done by studying the texts reflexively and socially and agreeing how texts might (if these categories were applied) be fitted. Therefore, after many texts had been looked at in this way, it was seen that all logically possible events could be classified like this, and that therefore, this would be an effective categorisation system. Now, when this tripartite distinction had been agreed upon (and codified), then it could be 'cross-referenced' with the 'frontline, supervisory, managerial' distinction discussed above, in order to create a matrix as follows.

This 'matrix' is a 'map' of the organisation of the industry. It is a sociological (*not* 'cognitive') 'model' of all activities within the industry. It was created for this specific analysis and socially decided upon, as was the meaning of each individual code within each 'cell' of the matrix, which were again created in an 'oppositional' sense derived from linguistics as described above. So for example, the Communication cell at the frontline consisted of two codes: communication between staff and communication from staff to supervisor/management. Each of these was defined in terms of each other (so that if a communcation error was between staff, then it could not have been from staff to supervisors, and vice versa).

The fact that a 'sociological' matrix was chosen derives from work done by Reason and others, who emphasise that, increasingly, an organisational and systems viewpoint is proving to be an effective approach to the analysis of safety issues in organisations (Reason 1997. See also Shrivastava *et al.* 1991; Hutter 2001).

So we assign a maximum of three codes within each 'cell' of the matrix (for adequate discrimination), allowing for a maximum of 27 individual human factors codes,[5] each of which can be assigned a numeric value, and thus transferred to the statistical database. At this level, therefore, a piece of text, having been 'taken apart' and 'reconstituted' was then 'coded'. 'Medium-sized pieces of text' were assigned a place in the 'matrix': for example, if a supervisor failed to communicate something to staff then it was 'coded' with a 'coding stripe' at the supervisory level, then at the communication category. Then, given that the text stated that it was a failure to communicate to staff, this 'code' was applied. And how was this decision made? As Ricoeur states, it is a decision of 'subjective probability': is it more or less likely that this particular piece of text fits in this particular part of the matrix, and fits this particular 'code'? (Ricoeur 1977: 331).

There are a number of points that should be noted here. The first is this methodology ensures that readers must stay true to the 'hermeneutic circle'. First there was the initial reading in which subjective intuition as to the transcript's meaning and purpose was made (this was performed at the point of the initial reading of the short description in the Access database). Then the whole transcript was broken down into 'micro-units' of meaning, which were then related to the meaning of the whole again (considered in the light of the frontline, supervisory, managerial distinction), before being related back to individual segments this time in such a way that fragments of text can be assigned 'codes'. This concept is of course taken from Heidegger, but Ricoeur has adapted it to his own approach (Ricoeur 1977).

Secondly, the methodology was derived from Ricoeur's hermeneutic arc. Therefore the basic distinction into 'how', 'who' and 'where' (etc.) correspond to his initial hierarchical classification, whereas the reading via the matrix in Figure 6.2 reveals the text's deep structure. The matrix and codes were a way of revealing the deeper meanings relevant to safety which might lie underneath the surface linguistic structure; fragments of text were then selected and coded. Needless to say, the codes themselves were developed from the matrix but were more precise in describing specific fragments of text insofar as they threw light on these underlying meanings.

	Job/Task	Communication	Rules/Procedures
Managerial	Codes	Codes	Codes
Supervisory	Codes	Codes	Codes
Frontline	Codes	Codes	Codes

Figure 6.2 Matrix.

So the process was hermeneutic. First, coders were trained with a codified methodology in a shared social context in an explicit attempt to create *shared* 'traditions' and 'prejudices' in terms of an approach to the texts. The actual approach to initial texts by a trained coder began with an initial 'subjective' 'guess' made from the data in the Access database. Once the hermeneutic circle had been entered, the whole (interview) text was then read through. This text was then broken down into surface elements (who, what, where, etc.). It was then reconstituted and read as a whole again before being broken down into elements classified via the matrix (Figure 6.2). Finally it was reconstituted, read again, and produced as a graphical representation, before being broken down for the third time into discrete textual elements which were then assigned codes.

It must be stressed that codings were not either 'correct' or 'wrong'. Instead, as the system is developed, a process of 'juridical reasoning' which has a 'polemical character' was entered into (Ricoeur 1977: 332), at the level of, first, the basic classifications, then the basic codings applied. Nothing was ever set in stone, and individual coders could always argue that codes needed to be changed or upgraded, or that other people's codings were more or less appropriate in a given context.

Verification

This still left the problem of verifying the various readings produced. This was particularly important with a hermeneutic approach given that in the hermeneutics and functional discourse tradition in which we were working, meanings are not fixed by internal cognitive mechanisms but are instead produced and read in a dynamic social context (Winograd and Flores 1986). Therefore, we could not simply assume that because we were 'experts' in psychology we could gain privileged access to 'what the reporter really meant'.

So coders did not pretend to know more about railway safety than the people who produced the original texts. However, it had to be demonstrated that there was sufficient concordance between interpretations such that coders meet the criterion of consensus (see Chapter eight). Therefore, reliability trials (or to be more precise, trials of concordance or agreement between coders) were carried out, which test the extent to which, using the hermeneutic method detailed above, we could interpret the same texts in the same way. Because the data produced could be statistically analysed this was easy to do.

It must be stressed that reliability is *absolutely essential* for testing the degree of concordance between interpretations especially in a field as important as safety. As has been argued before researchers must be able to demonstrate reliability or else effective functioning of a taxonomy is *impossible* (Wallace *et al.* 2002).

The reliability trial

The most important aspects of the applied hermeneutic reliability trial are that it was carried out under strict conditions (two independent coders, who were not able to consult with each other) and that the trial measured raw agreement (Wallace *et al.* 2002). The two coders were in two separate rooms. They were not allowed to discuss any aspect of the trial before hand. There was no discussion or communication of any sort during the trial.

This is particularly important in that it might be decided that individual reports could be said to contain specific safety events or concerns. For example one report might be decided by a coder to contain an incident (that is a safety situation which had a consequence) and an issue (a safety situation which did not have a consequence). The decision as to whether or not there was 'more' than one issue or incident in a report, and if so, whether it was to be classed as an issue or incident was part of the reliability trial. Subsections of the reports were not 'pre-isolated'. We recommend that this procedure becomes standard in other reliability (consensus) trials. This is particularly important in that the text on which the trial was carried out was the natural language of the interviewee: it was not re-written for the purposes of the trial.

The results of the trial are shown in Table 6.1. Totals were calculated using the Index of Concordance: A/A+D. The overall total between the two coders was therefore 76.54 per cent. In a standard work on research, Borg and Gall state that reliability of over 70 per cent is acceptable in this context (Borg and Gall 1989). By this criterion the system of applied hermeneutics used here has clearly been shown to be reliable.

Kinds of data

The most obvious question now is, what kind of data can be produced from this method? We would say that for people attempting to manage safety issues in a specific industry or organisation, there are two main answers to this question: snapshots and trending.

Snapshots

A snapshot is what it sounds like; a description of the state of the discourses at a specific time. So all the 'codes' are added up, and the total number of codes assigned at each level at one moment in time is shown. The graph in Figure 6.3 is therefore a fictitious representation of 'problem types' (i.e. codes assigned) at the managerial level. They are presented as they are grouped in the matrix.

Table 6.1 The consensus (IRC) trial: agreement between coders A and B (%)

Event reports									
1	2	3	4	5	6	7	8	9	10
55.55	78.57	72.72	68.75	100	87.5	66.66	75	73.3	87.5

Trending

Trending is simply a way of showing the codes in a temporal manner. The graph in Figure 6.4 is a fictitious representation of all job/task, communication and procedural issues across all levels across the time period of the project. This demonstrates visually the relationship between these three factors, and obviously shows which features are 'improving' and which are 'getting worse' at different points in the time frame. Though such a representation is sometimes referred to as 'trending', a precise definition requires a statistical analysis to establish whether a trend is significant, as opposed to mere visual inspection.

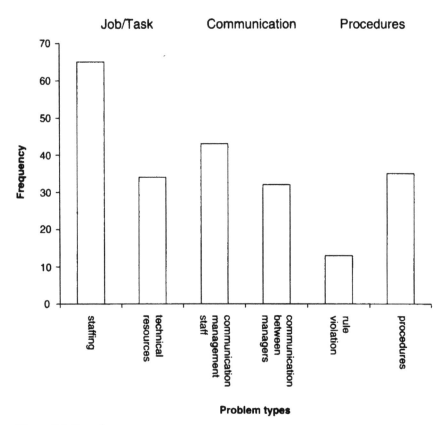

Figure 6.3 Snapshot.

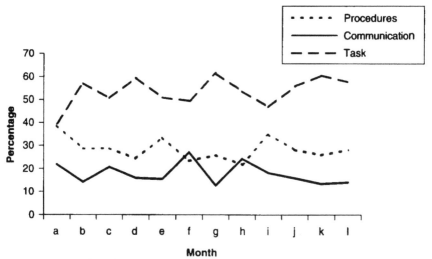

Figure 6.4 Trending data.

Control charts

The above information is not meaningful unless one has a baseline criterion for deciding how 'bad' is 'bad enough', that is what amount of the fluctuations in Figure 6.4 is merely random statistical background 'noise' which might be expected anyway, and what amount indicates definite safety issues which must be dealt with. Therefore we sought the help of the Statistics and Modelling Science Department at the University of Strathclyde for assistance in modelling the data. All data of this type (that is, quantitative data) which vary over time can be modelled statistically (in this case) as Poisson or Binomial data, that is data which follow the Poisson or Binomial distribution pattern. Both of these can be used, not just to model data but in order to define levels within which the data vary due to chance. So for example, if one takes the Poisson data for a specific variable one can say that due to random chance it should not exceed a present parameter more than once every one hundred observations (weeks in the current system). In this example, the parameter is set at 5 per cent. The data were modelled (that is, the distribution was seen as being, for example, Poisson or Binomial), and, thus transformed by various statistical techniques, were represented as a number between 0 and 1.

We can see from Figure 6.5 that the variable 'attention' was 'out of control' during week 45 of this time series (this terminology is taken from engineering in the use of the 'control chart'). This indicates that this might be the time when further investigation might be necessary. Of course if the variable was to consistently breach the 'warning line' this might mean that further investigation of attention issues in this industry as a whole is necessary. The 'warning line' is set at '5 per cent' in this case study purely for

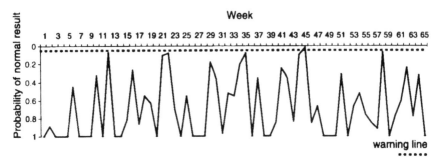

Figure 6.5 Control chart: attention.

purposes of illustration. (It may be set or reset at different levels in the light of experience or on the basis of expert judgement.) So variation of the data between 0.05 and 1 is 'normal' or 'in control', and between 0 and 0.05 is 'abnormal' or 'not controlled'.

It must be stressed that the control data above are reliable data produced by an established methodology which were then turned over to expert statisticians who analysed the patterns that exist in the data and applied controls to them. The use of engineering terms is therefore not metaphorical.

From hermeneutics to action

The previous paragraphs have talked at some length about language, philosophy, the hermeneutic cycle and a number of other things that seem rather esoteric and somewhat removed from action in the real world. However, there is nothing so practical as good theory, and practice that is unguided by any theory succeeds, if at all, through luck rather than coherence. Even if it 'works' it is often difficult to see why. A few comments are thus in order to demonstrate that the ideas of some rather remote philosophers have some very straightforward application, and serve as a rational basis for dealing with safety reports.

The first thing to say is that the system described is not rocket science. Like everything else, it seems strange at first, but once one is into the swing of things it takes longer to describe than to do. So what are the main features of this approach?

First, a safety report (whether a written report or a transcript of a conversation) is an object. It is a thing; it exists. Safety reports are produced by organisations and so therefore we can say that safety reports are concrete outputs from organisational systems. They are identical therefore, *in that respect*, to the output of sausages from a sausage factory, motor cars from a car factory, or power output from a nuclear reactor, apart from the fact that they are not the *primary* output of a system. But they are still objective outputs from a system. A safety report, as an object, is therefore 'objective' rather than 'subjective'. As objects, safety reports can be counted at different times, different places, analysed for trend and so forth. There is nothing interpretative (within the normal use of that word) with saying 'We received 30 safety reports last week' and viewed as objects there is nothing we can do with data about cars, sausages or power outputs that we cannot do with safety reports.

However, where safety reports differ from sausages, cars and megawatts is that before dealing with them as objective data they have to be interpreted, and the only way any piece of text has practical usefulness as a guide to collective action is if everyone involved in the collective action agrees about what it means. The hermeneutic approach suggested here is simply a way of obtaining agreement about what the important features of a safety report are, and the reasons why this is important will become obvious (if they are not already) in the next few paragraphs. Suffice to say that if the meaning is interpreted not merely subjectively, but differently, by different people, as described in the early chapters of this book, there is no possibility of focused action or of learning lessons, and safety proceeds piecemeal according to the different biases and predilections of individual interpreters (coders). There has to be reliability of coding or safety simply cannot proceed on a coherent basis. That is why an approach of the type suggested becomes a key feature of learning from event reports. Sausages, cars and power do not suffer from the problems of interpretation, not because they are more objective, but simply because within our culture everyone agrees a car is a car, and a sausage is a sausage, and is more or less in agreement about their functions. The world would be strange if they did not.

With safety reports, however, people have their own biases, theories about what a report 'really' means, preferences for certain types of causes rather than others and so forth. To learn lessons, therefore, there has to be an approach that allows people to agree, or even gently draws them to agreement by virtue of positive and hierarchically organised demand characteristics of a taxonomy, as to what the content and meaning of a report is whilst not restricting their choices and still allowing them to interpret. In other words, towards a hermeneutic consensus. Once this has been achieved, however, safety reports are just data points like any other data points.

At this point, the argument moves to the next step. Any systems output of whatever type varies over time; it is not perfectly constant. Sometimes this variation is marked, for example the wind speed might vary from zero m.p.h. to 30 m.p.h. without being seen as 'abnormal' or stormy. In other cases, the variation is much smaller; for example if I engage cruise control in my car for 60 m.p.h. it seems to vary only by about 1 m.p.h. under normal road conditions. In the same way, the pressure inside a nuclear reactor vessel shows variations over time, but most of the time these variations are within a range of 'normal variation'. In all these situations, it is possible to differentiate 'normal variation' from 'abnormal variation'. It is also possible to draw 'control charts' to express the point at which one wishes to differentiate normal variation from abnormal variation. Finally, when the system exceeds the limits of normal variation, one takes note of that fact and investigates the situation with the intention of fixing it. Now, once we have agreement as to how an event report is to be coded, we can treat the resulting data in exactly the same way as wind speed, speed of a car, or output from a nuclear reactor.

First, there is normal variation. Within any safety reporting system, there will be a background 'hum' of safety issues which represents normal reporting rates for various kinds of things. No-one is happy all the time; somewhere, some people are going to feel tired, unhappy with their shift, discontented with pay and conditions, dissatisfied with the design of a machine and so forth. They will submit reports about these things in varying numbers over time. In our experience, safety managers often refer to these things as 'moans' or 'gripes', though we prefer to call them background noise. The problem, of course, is to differentiate this background noise from abnormal variation; a system of the type proposed should be able to do exactly that once there is agreement on the meaning (content) of the reports. Once this is achieved an accumulating database of reports can be described in terms of control charts for any reported human factors issue, using exactly the same mathematics as modelling the output of a nuclear reactor. It becomes possible to say, for example, 'Reports of fatigue between 3 and 4 a.m. in location 'x' have been out of control for such and such a period of time. It's time someone went and had a look at what is going on there'. And that, of course, is exactly what one does with engineering or 'hard' data. In summary, therefore, the system described seeks to turn 'subjective' reports into the same kinds of data as the output from any other system, engineered or otherwise. It turns conversations into numerical safety data and differentiates abnormal reports from background noise.

The most important thing that we feel has been achieved via this process is that a unified approach to qualitative data has been developed that has some sort of 'check' on interpretations in the form of a reliability (consensus) trial. Moreover, this process is independent of speculations about whatever unverifiable 'cognitive' 'inner states' were going on in the subject's mind at the time of the interview. What we have analysed instead are texts considered as texts, using a matrix which illustrates non-causal, systemic relationships between aspects of those texts. The method of demonstrating reliability was through 'triangulation' of the perspectival approaches of readers. In other words it was non-cognitivist, non-causal and based on a pragmatic, (social) consensus view of interpretation. This fits in with the paradigm we have proposed in other chapters in this book.

In short, we hope we have demonstrated that qualitative and quantitative methods are complementary. There is no real contradiction between them. The only real contradiction is between methods which are reliable, accurate and precise, and those which are not.

7 Causal attribution and safety management

In the last chapter, we examined the need for consensus in the interpretation of event reports, and suggested a method based on the 'hermeneutic circle' for achieving this. Now it is time to look in more detail at the contents of the reports themselves, and at some of the features of reports that make their interpretation a little tricky. One of these problem areas concerns the ways in which people ascribe causes to things, and the way people do this is the major focus of a body of research and theory known as attribution theory.

It has been recognised that the question 'does causation exist' is not a simple one (see Lakoff and Johnson 1999: 232, also Chapter two of this text). Whereas nuclear reactors, trains and people 'undoubtedly' exist, the dynamic relationship which we 'see' between them when we describe them as 'cause' and 'effect' *cannot be directly observed*.[1] The word 'cause' in the statement *'the broken valve was the cause of the leak'* clearly does not denote a material object (like the valve) or an observable event (like the leak). The word is used to *link* these phenomena in a way that makes sense in a certain context, and this linkage is dependent upon the information available to the person making the statement.

It can be argued that, in an open system such as a major organisation, there are an infinite number of events that can theoretically be linked (i.e. said to have a 'causal relationship') with an accident (e.g. Toft 1996: 103). The causal links *we choose to make* when investigating accidents depend upon certain biases, which are outlined in this chapter. When something happens, we observe events and *state in language* that certain ones are causal. We construct a plausible account in line with available information and our own characteristics as investigators (see, for example, Dejoy 1994). In other words, our reason for attributing causality to events is primarily pragmatic, and an expressed 'cause' of an accident cannot be viewed as a physical or 'objective' feature of the world. 'Cause' is a word that can only be understood in a social and linguistic context. As Lakoff and Johnson put it:

> ... the concepts of cause and event ... are fundamentally human concepts ... Their meanings have a rather impoverished literal aspect; instead, they are metaphorical in significant, ineliminable ways. (1999: 171)

The purpose of this chapter is to show that people assign causes to safety events in functional and variable ways. Importantly, recognition of this fact allows for an analysis of attributional ('cause-choosing') behaviour that can shed light on the systems in question. Understanding this bias in causal investigation is vital for interpreting and understanding event reports, so that resources can be targeted wisely. For example, Reason (1994) points out that a tendency to attribute 'blame' to the most proximal individual does not tend to lead to effective interventions.

As stated above, evidence for the bias inherent in causal investigation comes from the area of psychology known as 'attribution theory'. There are two broad approaches in attribution theory, which are principally distinguished by the theories of truth and language adopted by proponents. Accordingly, two views of 'cause' and 'attribution' exist in the field. First, 'traditional' social-cognitive psychology allows for cognitive representation and 'real' causes. Alternatively, 'discursive' psychology sees 'social language function' as the subject of interest, and sets aside ideas of 'mental models' and 'true causes' that exist outside language. Both positions are described briefly below, and the implications for a discursive approach to attributions in safety management are then discussed.

Traditional attribution theory

'Attribution theory' is a term that has come to denote a body of work 'on the perception of causation and the consequences of such perception' (Kelley and Michela 1980: 458). Specifically, attribution theorists attempt to describe (and therefore predict) the circumstances under which different people give different causal accounts for events, and/or use these accounts to predict and explain future behaviour.

Heider (1958) suggested that people strive to explain events by accumulating evidence and describing causal links. Subsequently, Kelley (1967) modelled the process by which variance in incoming information might lead to different attributions being made. Predicting the type of attributions likely to arise in specific situations usually involves classifying them along the dimensions proposed by Weiner (1974a). Weiner proposed a model in which there are two dimensions (stable/unstable [stability] and internal/external [locus]) by which attributions can be defined. Further dimensions of controllability (Forsyth and McMillan 1981; Weiner 1979) and globality (Seligman 1979) have since been proposed.[2]

Attribution and mental representation

Traditionally, attribution theorists have proposed that the attribution corresponds to a *representation* of causality in the mind. This traditional *cognitivist* epistemology arises from the work of Locke (1689), who outlined the position whereby an idea in the mind stands for, or *represents* its external object.

Social cognition is concerned with people's representations and hypotheses about other people. However, the basic representational position remains. For the cognitivist, language is viewed primarily as a vehicle to transport us to a person's inner world of representations (see Peschl and Riegler 1999 for a critique of this approach, also Chapter nine of this text). For example, when Heider (1958: 297–8) concluded that people generate concepts about causality, he proposed they 'form the content of the cognitive matrix that underlies our interpretations of other people's behaviour and our attempts to influence it'.

Traditional attribution work follows from this assertion by Heider. Kelley and Weiner both subscribed to cognitivist principles to a certain extent, and saw attributions as corresponding to underlying mental states (e.g. 'causal schema', Kelley 1967).

Attribution as a reflection of external causality

Those who propose that attributions reflect inner representations have also traditionally viewed them as reflective of 'causes' which exist in the external world. Much experimental social psychology (e.g. Lalljee 1996) is still characterised by this view, in which the world of Newtonian cause and effect is the fundamental reality. The associated 'correspondence' theory of truth holds that a proposition is true if and only if the language corresponds to the relevant objective facts (e.g. Devitt 1984). So in causal attribution the attribution is true if it matches (*corresponds to*) the 'real' cause.

An important by-product of this view of language is that, when attributions are shown to vary, (i.e. they are *biased*), the concept of 'attribution error' (Ross 1977) is introduced to reconcile such variation with the assumption that causes are either 'true or false'. In this paradigm, people attribute causality, and because 'scientific' causality can be determined, people can be 'wrong'. Implicit here is a distinction between lay (common sense) and reified (scientific) causality (i.e. people assume *attribution* is one thing and *cause* is another). Problems with assuming that the latter (objective and mechanistic) can be distinguished from the former (consensual, social and teleological) have been pointed out (e.g. Davies 1997).

However, there is good evidence that the idea of an internal cognitive representation linked to an external and 'real' (non-discursive) causality may be misconceived, and fails to do justice to the situation.

Functional discourse and attribution

Problems with the idea of causal representations

Neisser (1967: 5) asserted that cognitive representations 'surely exist' and defined cognitive psychology in its infancy as 'inventing hypothetical mechanisms'. Criticism of the idea that the word 'cause' corresponds to a 'representation of cause' often derives from the epistemological views of

Wittgenstein (1889–1951).[3] Wittgenstein outlined an alternative view of language, underpinning a second tradition in social psychology and attribution. Wittgenstein (1958) argued strongly against this assertion that there are 'surely' cognitive mechanisms which we must therefore attempt to describe. He argued that an analysis of language behaviour would divert us from any need to look for a 'stored thought' independent of our verbal expressions. He thus raised the possibility that a verbal act (like the act of attributing a cause to an event) might not be evidence for some 'mental representation' of the words used. If we view attributional language as functional behaviour, the cognitivist position becomes hard to sustain. Attributions cannot be viewed as descriptions of a 'causal schema' (Kelley 1967) which people relate in a disinterested manner.[4] The notion of a 'cause' as an external object that is described through language can be criticised from a similar epistemological position.

Critiquing the idea of causes as external objects

In Wittgenstein's later work he argues strongly against the 'pipeline' (or 'conduit') metaphor whereby language is viewed simply as a way in which facts about the world are transported from person to person. Wittgenstein emphasised a set of functional behaviours with their own criteria and logic, which he called 'language games'. As he famously stated: 'For a large class of cases – though not for all – in which we employ the word 'meaning' it can be defined thus: the meaning of a word is its use in the language' (Wittgenstein 1953: 43).

Thus describing the 'cause' of an event is fundamentally performing a (useful) action, and not 'merely' telling someone else about a property of the world. So, if an investigator states that 'X causes Y', then 'cause' means precisely a word that the investigator has used to link X and Y, no more, no less.

Like Wittgenstein, Austin (1962a) was concerned with the philosophically misconceived notion that language can be treated as an abstract referential system. He started to dispel this myth by identifying two types of statement. 'Constative' statements are descriptive and state things about the world. An example would be *the train derailed*. 'Performative' statements are those that perform actions, for example, *I hereby declare that I am bankrupt*.

But Austin did not stop at establishing that some statements perform actions. He went on to form a general theory of 'speech acts' that included *all statements*. For example, suppose the statement above, *the train derailed*, is rewritten as *I hereby declare that the train derailed*. Now the statement is seen to be performing an action. The myth that statements can be entirely descriptive (i.e. they can have no performative or functional aspect at all) is thus exposed. All statements are functional; they are explanations formulated with their purpose or consequences in mind. As Austin himself puts it, 'Stating, describing ... are just two names among a very great many others for [speech] acts; they have no unique position' (1962a: 148–9).

Attributing causes to events can be added to Austin's list of speech acts. According to the later work of Wittgenstein (1953) and Austin (1962a), causal statements cannot be viewed as objective descriptions (flawed or otherwise) of external events or causes. They are statements that have a purpose, and making them is a kind of (language) behaviour.

The principles upon which the concept of a 'cause' is denied any special objective (as opposed to discursive) status were established by Hume (1739), who called a cause a 'union in the imagination' (see Chapter two). Others have since provided evidence for this position. For example, Nisbett and Wilson (1977) studied social decision making and provided evidence for the inappropriateness of viewing language as corresponding to objective causality. They showed that people gave cognitive-motivational explanations (attributions) for 'choosing' favourite clothes from a rack. However there was a strong tendency to 'choose' garments placed in a certain position, i.e. choice was shown to be *independent* of attribution. The 'language game' of attributing causes to the clothes chosen could be seen to be primarily functional. It did not reflect a 'real' reason underlying the choice.[5]

Similarly, in a series of experiments, Heider and Simmel (1944) presented subjects with moving shapes which they manipulated so that one shape would move followed closely by another. Occasionally they would come into contact. People were seen to ascribe causes and attribute social motives to the shapes. For example one shape would be described as 'chasing' another which would be 'trying to get away'. Similarly, Michotte (1946) conducted a series of experiments on phenomenal causality, in which he manipulated visual stimuli (shaped projections on a screen) so that they appeared to move in certain ways. He then investigated the circumstances under which people would make causal statements about the movement of the shapes. His central finding was that people would attribute causality (for example the movement of one shape being 'caused' by collision with another) to events where no such 'scientific' cause and effect existed. So people observed phenomena and associated these in causal language where it seemed appropriate. Further this was predictable: if the shapes were close enough together at the 'point of impact', then causality was ascribed. Thus the *phenomenological* appearance of causality could be removed by increasing the distance between the shapes at this point.

The results of these studies support the view that statements about causality are social, functional and motivated, and that causes assigned to events can be predicted given some knowledge of the phenomena involved. We will now discuss attribution in a safety management context, and describe how some of the functions and motivations inherent in the process lead to tendencies to attribute certain types of causes.

Causal investigation of accidents viewed as a functional act

Tendency 1 Ascribing causes that are similar to effects

The principle of *similarity* assumes properties of causes to be similar to properties of effects with which they are associated (McCauley and Jacques 1979). So for example, if the event is seen as being simple, people will look for simple causes. By this principle, the type of cause likely to be assigned to any given effect can be anticipated. Examples of this effect can be found in the realm of safety management. Investigations into a slight injury to a single operator are generally ignored by the media and have little social impact. Thus causes assigned may be proximal and relatively mundane, along the lines of 's/he tripped and fell'. In contrast, a train crash where members of the public are killed affects more people and has more social significance. The event here is in effect defined by its social consequence. Therefore, it is not likely that 'they were hit by parts of the train' will suffice as an attribution for the deaths of passengers. The explanation will generally derive from a detailed investigation and the causal 'chain' constructed will be more complex (i.e. involve more causes), involve more people and have more social impact. Note that principles of similarity are built in to investigation techniques. There is usually a statutory requirement to investigate (i.e. to attribute causality) which is contingent upon the social significance or consequence of the event in question. Thus the 'bigger' the event, the 'bigger' the investigation, leading to similarity in the features of the cause(s) and event. Put simply, attributions to managerial factors, for example, do not arise until events *seem to require* such explanations. It cannot be assumed that the 'causes' of major and minor events are inherently different just because investigators 'dig a bit deeper' into the system when it seems appropriate.

For example, at 3:30 a.m. on May 6, 1935, TWA Flight 6, a Douglas DC-2 flying from Albuquerque, New Mexico, to Kansas City, crashed sixteen miles south of Kirksville, Missouri. There were five fatalities. One of the fatally injured was Senator Bronson Cutting of New Mexico (a prominent politician of the time). Komons (1984) notes that :

> ... the Cutting crash was seen by many people as a tragic consequence of a bankrupt aviation policy – a policy, it was held, that neglected Government's responsibilities in air safety in favor of economy and political preferment (1984: 2)

So the outcome (death of a member of Congress) is matched to the attribution (Government policy). Others have noted how the causes assigned to the crash were not those of scientific determinism but were functional and social. For example, Rimson (1998) reports how 'the discord which followed the Cutting Air Crash permanently dashed any expectations that air safety investigation authorities could determine causes of accidents objectively without the self-serving intervention of implicated parties'. So Rimson points out that the death of the senator meant 'objective' investigation was impossible. However, it can be argued that a crash with *no* senator on board has demand characteristics for investigators as well (though they may be very different, i.e. get this thing mopped up and out the way). *Every* event has properties that affect the subjective accounting by investigators.

It should be noted that subjectivity in accident investigation is not universally acknowledged. For example, in investigating the Clapham Junction rail accident, Anthony Hidden QC is at pains to point out that 'An inquiry under the Regulation of Railways Act 1871 is not a trial: ... it is an investigation with the object of discovering the truth' (1989: 147). Subsequently, one such 'truth' presented in the report is that 'At the centre of the problems which caused the Clapham Junction accident were the bad wiring practices followed by the workforce in the S & T department and allowed to continue unchecked by its management' (160). It can be seen that the epistemological status of the 'causal' statement is supposedly established by the first, 'truth' statement. This is similar to the process in a court of law, where a witness statement 'I will tell the truth' is used to imply objectivity in another, for example, 'it wasn't me, it was him'. People are quick to spot the functionality of such pleading of innocence, but Hidden explicitly adheres to the assumption that discourse corresponds to an external reality, and people giving evidence have merely described their representations as to what happened. Wetherell and Potter (1988) call this an 'old-fashioned' view of language whereby it 'acts as a neutral, transparent medium between the social actor and the world, so that normally discourse can be taken at face value as a simple description of a mental state[6] or an event' (168).

Tendency 2 Ascribing causes that are closer in time and space to effects

The principle of spatial contiguity states that events closer in space to the target event will be more readily assigned as causes than those further away. As discussed above, Michotte (1946) showed that the phenomenological appearance of causality can be removed by increasing the distance between the shapes at this point. Similarly, the principal of temporal contiguity assumes that people will assign causes more readily if they appear at essentially the same time as (or just prior to) a given effect (Seigler and Liebert 1974). So, for example, a supervisor who forgets to pass on information to a new operator may be 'blamed' for a mistake the recruit makes *within a certain period of time* afterwards. The greater the time between the 'cause' and 'effect' (and thus the more events in between), the less the likelihood of the former being perceived as a cause of the latter. After the operator has performed the task correctly once or twice, s/he will not be new anymore, and mistakes will be more readily attributed to her/him.

These associated principals of spatial and temporal contiguity provide a useful way of looking at certain biases in safety management based on 'root causes'. Blumenthal (1970) points out that events closer in time and space to an accident are most likely to be assigned as causes, often based on superficial observation. In our own research in the nuclear field (Ross *et al.* 1999; Wallace *et al.* 2002), regularly occurring clusters of root causes (as applied by safety managers) relating to a) technical factors and b) 'work practice' human factors at the level of the man–machine interface were identified.

This finding is borne out by other research in the area of human factors in industrial systems. First, Wilpert and Fahlbruch (1996) note the tendency of those involved in event analyses to 'focus on technical component features at the expense of the human contribution'. Second, a bias in event investigation towards 'active' as opposed to 'latent' failures (for example in respect to aircraft accidents) has long been recognised (Reason 1990). It is axiomatic that either technical items or frontline staff (or usually both) are *spatially* and *temporally* close to accidents. However, this is not always true of supervisors, designers, maintenance crews, selection and rostering managers, writers of procedures, etc. Thus a bias in causal attribution to technical failure and staff at the 'sharp end' is predictable in theoretical terms, and should not be taken as *prima facie* evidence that these factors 'cause' more accidents than the abstract factors associated with other (usually more senior) staff. The common view that 'human error' accounts for a substantial proportion of accidents (e.g. 90 per cent: Larson and Merritt 1991) must be observed with the proviso that 'human error' is *predictably the most likely causal attribution* in the first place.

Tendency 3 Ascribing causes to salient features

In addition people are thought to attribute effects to causes that are more salient in the perceptual field. For example Taylor and Fiske (1975) varied the seating arrangement of actors in a film so observers saw different actors 'front-on' when viewing the film. The groups of observers were shown to differ in that they assigned a more *defining role* in the interaction to the actor they could see more clearly. In a well known study McArthur and Post (1977) manipulated features of actors involved in an interaction. For example one actor would be more brightly illuminated than another. Subjects viewed the interactions and were asked to describe what they saw. Results showed subjects attributed behaviour to more dispositional factors in the outstanding actor.

Once more the bias towards 'human error' and technical 'causes' in accident investigation is consistent with this principal. One might, for example, argue that frontline staff, in their bright, clearly identifiable company uniforms, are perceived as being more salient than the middle manager in his grey pinstripe suit. Similarly, leaks, breaks and corrosions are seen as salient in a way that cultural barriers to adequate communication between maintenance staff are not. Thus, bias towards the former is expected, regardless of difficulties in establishing consensus in the latter case.

Tendency 4 Ascribing causes based on first impressions

Research also shows that information assimilated first is more likely to be used in causal accounts than that gathered later. This is known as the principle of primacy, which in essence stresses the robustness of first impressions. Thus a person 'scans and interprets a sequence of information until he attains an attribution from it and then disregards later information or assimilates it to his earlier impressions' (Kelly and Michela 1980: 467). Evidence for this effect comes from Jones *et al.* (1968) who asked observers to judge the ability of people whose task performances varied over time, *but were ultimately matched.* Higher ability was attributed to those who started well (and tailed off) rather than those who started badly (and got better).

There are many examples of this effect in attributing causes to safety events. On the day of the Ladbroke Grove rail crash in October 1999, it was immediately reported that driver Hodder (who tragically died in the crash) had passed signal SN 109 at danger. Within a matter of days it was widely noted in the press that he had been convicted of assault in 1998 (for a minor offence in respect of which he was given a conditional discharge). The clear motive for such reporting was to encourage the notion that the tragic events might be attributed to driver Hodder's personal characteristics. What price a front page article on the personal social histories of the Thames Trains managers, those responsible for reviewing signal SN 109 (which had been passed at danger on eight previous occasions); the signallers; the consultants involved in reviewing implementation of Automatic Train Protection in the Thames Trains fleet? It may be argued that, in such cases, no amount of contributory system factors, discussed in subsequent inquiries, eroded a *primary impression* that blame lay with the driver. How many people were still interested by the time Counsel to the Ladbroke Grove inquiry, Robert Owen QC, concluded that driver Hodder's criminal record 'does not appear to have any bearing on the causes of the collision'?

Commenting on the aftermath of the Cutting crash (see above), the president of the Air Line Pilots Association detected a tendency, in the absence of direct evidence, for accidents to be conveniently blamed on the pilots involved. The pilot or driver, spatially and temporally contiguous, perceptually salient and primarily involved, would seem to have little chance of avoiding the strong pull of the attributional current (Perrow 1986). However, actions targeted at such individuals may not have the desired effect (Reason 1994), because the results of the investigation are essentially driven by features of the investigation process. (Of course, 'attribution bias' also affects academics working in the 'human error' field, whose tendency is to attribute events to internal, cognitive states of operators. Operators involved in such events, however, prefer situational, environmental and social explanations for the same events.)

Tendency 5   The self-serving bias

The evidence on similarity, contiguity and salience shows that differences in available information affect the causes a person will assign. Similarly, research has shown that *different people* will assign different causes given the same information. For example, it has long been known that investigator background affects causes 'determined' in accident investigation (Lewycky 1986). In outlining their 'Theory of Correspondent Inference', Jones and Davis (1965) argue that the antecedents of the attribution process are dispositional as well as environmental. A number of effects have been demonstrated. However most fall under an umbrella which has come to be known as the 'self-serving bias' (e.g. Miller and Ross 1975).

Examples of the self-serving bias can be found in the safety management field. In a piece of research with a major train operator (commercial in confidence), discussion groups were held with drivers who had passed a signal at danger and those who had not, in order to establish risk factors (driver attributions) for Signals Passed at Danger (SPADs). Drivers with no SPAD on their driving record attributed these unwanted events to more dispositional factors (i.e. things to do with the drivers, such as inability to concentrate) than the drivers who had previous SPADs. By contrast, this latter group listed system and environmental conditions (e.g. training, railhead conditions) for these events in general, and for their own incidents, thus implying the SPADs were 'caused' by external factors beyond their control. Thus 'causes' of SPADs seemed to be presented in a functional manner and showed a self-serving bias, dependent on drivers' own personal histories. Investigations into 'causes' of such incidents must take these presentational effects into account.

Of course, functional attribution is not exclusive to frontline staff. During the course of investigations into the Cutting Crash, The Bureau of Air Commerce's records of 'probable cause' determinations' in air carrier accidents were reviewed. As Komons (1984: 11) puts it: 'No-one was very surprised when the Sub-committee's Report proved to be remarkable for its disregard for factual accuracy, its misinterpretation of events, and its unrestrained bias ...'. An aide to the Senate sub-committee commented that '... in no case during the past ten years has the Department ever accused themselves of an accident' (10).

Self-serving biases can lead to differences in attributions between groups. For example, Hofmann and Stetzer (1998) studied attributional tendency in 'workers' and supervisors, and found differences in attribution for worker accidents, with supervisors tending to assign dispositional causes, and workers linking accidents to situational aspects. So the workers' tendency was not to blame themselves, and the supervisors tendency was to stick to proximal (worker) explanations, which can be assumed to serve a purpose in that it directs criticism away from organisational (i.e. management) failure (see, for example, Lacroix and Dejoy 1989). Similarly, Lehane and

Stubbs (2001) found low agreement[8] between accident subjects and their managers in attributing causal responsibility in 'slips and trips'.

To sum up, we must be careful not to see attributional tendencies as *errors*. 'Experts' do not have access to a 'real' scientific causality which can be compared against 'causes' assigned by 'ordinary mortals'. Rather, everyone tries to make sense of the world in a subjective manner. What then are the implications of this large body of evidence on attributional tendencies and subjectivity? How do safety managers go about applying such evidence in safety management? The answer is simple: instead of looking 'through' language to 'real' causes or objects the safety manager must look at the language itself. An example is given in terms of the phrase 'safety culture'.

Attribution and safety climate/culture

Safety culture: an objective feature of an organisation?

Despite common usage, there is no agreed definition of 'safety culture'. Indeed Cummings and Worley (1997: 479) state that 'Despite the increased attention and research devoted to corporate culture, there is still some confusion about what the term *culture* really means when applied to organisations' (*emphasis as original*).

Rousseau (1988) distinguishes safety climate (the sum of the individual perceptions of the organisation) from safety culture (the expression of shared or group beliefs and values). However, Rousseau points out that the distinction is a difficult one to use in practice as there is considerable overlap between the two. Sutherland *et al.* (2000) use the terms interchangeably because of difficulties in this regard. The International Nuclear Safety Advisory Group (INSAG) of the International Atomic Energy Agency (IAEA) define safety culture as 'that assembly of characteristics and attitudes in organisations and individuals which establishes that, as an overriding priority, nuclear plant safety issues receive the attention warranted by their significance' (INSAG 1991: 4). Pidgeon (1991) includes beliefs, norms, attitudes, roles and practices in a wider definition – further definitions are provided by Brown, Willis and Prussia (2000). Without listing all available definitions, a certain lack of consensus as to what 'safety culture' means emerges.

Where attempts have been made to produce reliable methods for assessing safety culture these have usually involved questionnaire methods. The 'competing values' approach (Denison and Spreitzer 1991) is typical in that it relies on a survey designed specifically for cultural assessment. Sutherland *et al.* (2000) assess safety culture/climate by determining attitudes to safety using a representative sample of employees, and a questionnaire designed for the purpose – further examples include the HSE Climate Survey Tool (1998); TRIPOD Delta condition survey (Groeneweg 1998).

Administering a 'safety-climate' device, of course, requires an assumption that people respond in a disinterested, unmotivated fashion and that results obtained represent a means of finding out the 'truth' about culture. However, it has been shown that 'the answers people give serve important functions for that person (e.g. self-presentation, preservation of self-esteem, apportioning credit or blame)' (Davies 1997: 83). There is considerable evidence that people can 'read' such experimental situations giving rise to artefacts that may be misinterpreted as veridical statements about pre-existing attitudes. For example, it can been shown that varying the person administering a questionnaire leads to substantial variance in answers given (Davies and Baker 1987).

Most questionnaires involve questions (or 'items') with a range of possible answers (usually five or seven: Likert 1932). A common tactic is to ask people how much they agree with a statement and have a range of options from *strongly agree* through to *strongly disagree*. Sometimes, these answers are treated as 'nominal' or categorical. In this case, a researcher might simply report that five people *strongly agreed*, sixteen *agreed*, twelve were *unsure*, five *disagreed* and ten *strongly disagreed*.

In most cases, the response options are treated as rank ordered. For instance, five people would be reported as ranking the statement as *agree* and ten as ranking it as *strongly agree*. Here, a set order to the items is assumed, where *agree* is *presumed* to be 'below' *strongly agree* but 'above' *disagree* and so on.

Sometimes, each answer (e.g. *strongly agree, agree, unsure, disagree, strongly disagree*) is treated as a point along a continuum which has equal intervals between each point. That is, *agree* is assumed to be to *strongly agree* what *disagree* is to *strongly disagree*. Similarly, *agree* is as far from *unsure* as *disagree* is from *unsure*. This position is attractive, as the numbers that answers correspond to can then be treated like any other 'interval' data, for example as if they lay along a slide rule. Means and standard deviations can then be reported for each question, and the numbers generated can be analysed using statistical techniques which take advantage of the numerical relationships between answers (see, for example, Mohamed 1999).

However, many have expressed concern about this assumption of equal differences between such rating categories. Put simply, it is probably not wise to assume that there can be 'objective' differences between ratings of agreement in the same way that there are between, for example, inches or seconds, or to justify parametric statistical analysis (e.g. t-test; regression) on this basis. Gephart (2001) describes a previous study where he showed this approach to be flawed. Similarly, Greene and D'Oliveira (1982: 26) caution that 'you should always consider what a conversion from an ordinal scale to an interval scale implies and whether it is appropriate'.

Using survey methodology for cultural assessment involves assuming people are capable of (and interested in) objectivity, and not motivated to answer in certain ways. It may be that difficulties in pinning down a clear definition in this area are borne out of this 'objectivist' assumption that 'safety culture' is an *object* which people can tell us about. Surveying for 'safety climate' by asking people 'what is the safety climate like round here?' assumes culture is a measurable object external to the language in which it arises. This then leads us to try and access this 'thing' through language. When this proves difficult, alternatives are proposed until a number of conflicting definitions emerge.

This is analogous to the view of attribution of 'causal factors' to be more than a linguistic tool, i.e as a reflection of some 'real world' property of a system. However, it has been argued that this view is implausible when examined in light of work from sociology, philosophy and literary theory which demonstrates the essential and inescapable 'action orientation' (Heritage 1984) of discourse.

An alternative definition can be given that emerges from viewing 'culture' as similar to 'cause'; i.e. as a linguistic phenomenon.

A discursive definition of safety culture

If it is not accepted that 'safety culture' is an 'external feature' of the world (i.e. the organisation) which people can 'tell us about' (see discussion above on the meaning of language), then 'culture' must be viewed as situated in (and constructed by) the way people talk. Defining 'culture' as *situated in* language rather than *perceived through* language has the advantage of using observable bias in people's accounts directly for measurement purposes, rather than viewing these as 'errors' which obscure the 'real' data, which will always beg the question of *how to access* the latter.

It can be argued that, as tendencies to attribute causes in various ways are functional, they may be a good starting point for a definition of safety culture that avoids unwarranted assumptions about language and can shed light on systems.

Suppose we view 'organisational culture' as *exactly equal to the attributional language of the organisation.* Now culture really is turned into a 'thing'. Language can be seen, listened to and observed in a way that is impossible with attitudes, beliefs, etc. The language is no longer the inappropriate measurement tool, but the object of measurement itself. We no longer have to *relate* 'culture' to the things people say (i.e. the causes they assign, see for example Hofmann and Stetzer 1998). Culture in this 'model' *becomes* the attributional language people use, and can be gathered, measured and analysed as we see fit. The following example is intended to show how a linguistic, attributional definition of culture might work in practice.

An attributional analysis of train drivers' explanations

Self-serving biases in company reporting systems were, implicitly, one of the reasons for the setting up of CIRAS, the U.K. Railway's Confidential Incident Reporting and Analysis System described elsewhere in this text. The system allows rail staff to report, in confidence, incidents and general safety concerns. They are then asked what the causes of these events and issues might be. An attributional analysis of some data collected during the developmental stages of this system, prior to it becoming national, is now described, which showed certain features of interest from an attributional perspective.

By way of introduction, it is worthwhile pointing to a well-known example of the clinical consequences of certain patterns of attribution as shown by Peterson *et al.* (1982) who report that 'depressive' symptoms are associated with an attributional style whereby events are attributed to stable, global and internal causes. In the classic sense 'internal' roughly equates to 'features of self'.

However, the interest here is of course the health of *organisations* rather than individuals. Accordingly, in this context 'internal' attributions were defined as attributions to self *or co-workers* (for example a train driver blaming drivers for events), and 'external' attributions were defined as attributions to *another group* in the organisation (usually managers). Further, it was agreed that organisations where groups do not attribute unwanted events *to themselves* are more at risk, because this shows a lack of collective willingness to accept responsibility.

Peterson *et al.*'s approach was adapted so that 'unwanted' organisational attributions were of the type; 'this is always a problem' (stable), 'it's a problem in many areas' (global) and 'somebody or something else is responsible for it' (external)'. It was posited that such attributions might be viewed as evidence of a problematic or 'unhealthy' 'safety culture'.

According to Abramson *et al.* (1980) it might be expected that where attribution is made in the terms described above, there will be an unwanted behavioural effect. In addition to monitoring attributions over time, systems such as CIRAS in principle allow for the cross-tabulation of types of causal assignation with other aspects in the discourse. Of particular interest are 'frontline' factors as experience with confidential systems suggests that around 75 per cent of reporters are frontline staff.

In one study (Ross 2003), reports where 'unhealthy' attributions emerged (i.e. where events were attributed to stable, global, external factors) were identified. All reports were then cross-tabulated with a number of 'frontline factors.' An example of such a cross-tabulation is given in Table 7.1, which shows the contingency table for *fatigue* issues and attributional pattern.

It can be seen from Table 7.1 that 111 cases of fatigue were identified from 1,157 reports (9.59 per cent). However, 22.8 per cent of reports where this style emerged raise concerns about staff fatigue, whereby only 5.2 per cent of cases without this pattern of attribution are associated with fatigue ($p < 0.001$).[9] Thus reports of fatigue seem to be associated with a 'can't do anything about it' style of explanation. The results, whilst not conclusive, suggest areas where a closer look might prove informative.

Chapter six in this text shows how a model of the general CIRAS discourse was developed so that distribution parameters could be established and

Table 7.1 Fatigue reports from drivers, by style of attribution (n = 1,157)

	Fatigue (%)	No fatigue (%)	Totals (%)
Stable, global, external attribution	66 (22.8)	224 (77.2)	290 (100)
No stable, global, external attribution	45 (5.2)	822 (94.8)	867 (100)
Total	111	1,046	1,157

control charts could be produced to aid industry decision making. Major accidents, policy changes, restructuring, are all factors which might in principle lead to an increase (or decrease) in certain types of attribution (i.e. in our definition *a change in safety culture*). Similar mathematical models might be developed in future to show which types of attribution increase in response to specific external events and which do not. This would allow for the cultural effects of such events to be evaluated.

Attributions and implications

The study of attributions (causal explanations)[10] has a long history in the social sciences, and the central aspects to this body of work have been briefly described, in order to show how causes assigned to safety events are best viewed as functional acts.

An attributional definition of 'safety culture' has been proposed which is *situated in* rather than *measured through* the language of people involved in an organisation.

This involves rejecting the view of causal language as reflective either of mental representations or external features of the world. In the first case, we discussed a distinction between attribution grounded in 'old-fashioned' social cognition, and attribution in modern, discursive psychology. Wittgenstein argued against the myth that meaning exists for an individual and is reflected or described through language and thus against the notion of a private world of mental objects correlated with a public lexicon of mental terms. Thus internal representations of the linguistic form of an attribution or event should not be assumed *a priori*.[11]

Similarly, compelling evidence exists that the causes people assign cannot be seen to correspond to external features of the world. This leads to an alternative, 'cybernetic' epistemology that stresses the social aspects to attributing causality. Here the results of an accident investigation become 'true' *if people agree that they are* (see Davies 1997: 58–66; Heylighen 1993).[12]

Despite the philosophical objections that have been highlighted, it has to be accepted that many safety 'experts' still *talk* about safety systems *as if* causes are properties of the system. Thus deterministic, 'causal' explanations are still the norm in event reports, investigation findings, etc. This begs the practical question of *what to do* with the 'root causes', 'causal chains' and 'contributory factors' in event reports and investigation findings. This chapter has attempted to show that:

- these must be primarily examined from an attributional perspective that takes their functionality into account;
- rigorous analysis of these explanations can shed light on systems, for example by providing a definition of 'safety culture' that has previously been elusive.

Examining accounts in terms of their function involves deciding which can be *agreed upon* and which are *variable*. Suppose a reactor trips. An investigator observes some numerical data and says:

1   *The reactor tripped at time X because the heat exceeded value Y.*

Another reactor trips. A supervisor is responsible for overseeing a valve. The investigator interviews the supervisor, observes the verbal data (*discourse*) and says:

2   *The reactor tripped because the supervisor was thinking about something else.*

The two causal statements are epistemologically identical. However, if there is seen to be a difference in the *consensus* with respect to these accounts (see Chapter eight), then it is possible to proceed by seeing a *practical usefulness* in one that may be absent in the other.

Suppose that the association between the reactor trip and the temperature rise is a common one, and most people attribute such trips to such rises. However, it may be that the supervisor does not stick to his original account, or that different investigators interviewing the supervisor would come up with different causes for the event. Statements 1 and 2 above can now be differentiated. One is primarily consensual, and useful in predicting reactor trips. The other is less consensual, less useful in terms of predicting trips, but useful in terms of the function the discourse performs for the person, which in turn may shed light on the organisational systems in question. A study of variables associated with *biased* accounts may provide information on investigators, investigations, supervisors, communications, etc.

This pragmatic approach allows the philosophical aspects of causality to become secondary. Some accounts are robust, and can be *best viewed* by setting their motivational aspects aside. However, this chapter shows how the attribution of causes to events involving human actions can most often be seen to be unreliable (functional and non-robust). Thus these accounts are best viewed from a pragmatic standpoint whereby the subject matter becomes the variability (or consensus) in the accounts themselves. This is especially true of 'causes' which involve attributions to 'mental states' or cognitions, because these are unobservable and hypothetical.

Finally, it is not particularly new to recognise the functional nature of language. The ideas of Wittgenstein and others are well developed in the wider scientific community. The epistemological arguments described in this chapter lead to a way of looking at the causes people assign which can broadly be described as *discursive* (e.g. Wetherell 1996).[13]

Potter (1999) describes a history of 'renewal from the outside' in psychology whereby traditions in linguistics, ethology and neuroscience have all provided impetus for new thinking in psychology. The particular impetus for a discursive psychology (e.g. Potter and Wetherell 1987: Edwards and Potter 1992) comes from the work of Wittgenstein and Austin described above, and the conversation analysis of Harvey Sacks (1992). According to Wetherell (2001) 'most discourse analysis proceeds through a questioning of simple realist assumptions that language is neutral and transparent' (392). Edwards and Potter (1992) state that 'everyday discourse has dynamics of its own which render it as evidence of underlying processes highly problematic' (31). Similarly, Kress (1990) argues for language as a social practice, i.e. as text produced by socially situated speakers or writers. Meanings come about through interactions between speakers and readers and are never arbitrary. Discursive function is the basis on which modern psychologists have become involved in 'critical discourse analysis' (e.g. Kress 1990; Chouliaraki and Fairclough 2001). The important issue of practical usefulness cannot be overstated.

It is important to avoid an approach which leads to a shrug of the shoulders and acceptance that discourse cannot be used to say anything about material objects. Many have grappled with the need to reconcile construction in language (even their own) and the need to say something about 'the world'. Some discourse researchers (Parker 1992) are happy to concede that there is a 'truth', albeit one which is always relative to some discursive, cultural or social frame of reference. However Edwards *et al.* (1995) argue in describing work on emotions that all we ever have access to is the discourse. It can be argued that Edwards takes the easy way out by using an example like emotion. It is relatively easy to adopt an *idealistic* position with respect to unobservable events like emotions (Edwards *et al.* 1995), causal attributions (Hume 1739) or cognitions (Skinner 1984). However, describing how the discursive and constructive relates to the *material* is more problematic. Wetherell asserts that ontological distinction between the discursive and the non-discursive makes investigation into the determination of one by the other possible, stating:

> critical discourse analysts tend to take a more materialist position indicating that they have an interest in a real material world independent of talk and discourse. (2001: 392)

So critical discourse analysis thus allows for the studying of discursive function and construction and the relation of this to *non-discursive* events (see also Chouliaraki and Fairclough 2001: 28).

These are difficult issues and the need for a consistent epistemology is vital. But some discursive events cannot become more 'real' than others. It cannot be assumed that statements about certain events (e.g. *the train hit the buffer-end*) are veridical, and other statements (e.g. *the driver was not paying attention*) are constructive and functional. All accounts are subjective or *biased*. The evidence in this chapter shows it is not useful to see some causal accounts as *wrong* and some as *right*. Post-event discourse must be seen as reflexive, and part of the system itself. Only by building analyses of bias into the investigation process can post-event data have any usefulness. This is why it has been argued that *consensus in accounts* is important as a means to establish any differential between these statements. In this way, functional rather than (unwarranted) epistemological differences between statements are established.

8   Inter-rater consensus in safety management

Introduction

Work in safety management often involves *classification* of events using coding schemes or 'taxonomies'. Often, such schemes contain separate categories, and users have to choose which one(s) apply to the events in question.[1] Taxonomies (especially those which are used 'after-the-fact') typically involve classification of features of a system (for example, human behaviours, organisational/environmental factors, or cognitive factors)[2] which can be examined to help avoid unwanted events in the future.

'First-generation' Human Reliability Assessment (HRA) approaches (see Hollnagel 1998: 122 for an exhaustive list) concentrated on operator behaviour whilst the recent trend has been towards classification of system factors (e.g. Reason 1990; Groeneweg 1992) and underlying cognitions (e.g. Hollnagel 1998).[3] Nevertheless, this shift in emphasis does not alter one crucial methodological aspect. If a coding system involving a choice between categories is to be useful, people have to be able to agree which categories to choose.

This chapter shows how use of the term 'reliability' to apply to coding schemes in safety management has led to confusion, due to conflicting definitions of the term. The 'traditional' definition in safety management is adopted from engineering. In this case, reliability roughly equates to *consistency* in comparison to an 'objective' standard. However, we will outline an alternative 'human science' definition of 'reliability'. Here, the crucial aspect is whether *subjective coding decisions* by different users lead to the same codes being applied (i.e. whether there is any *consensus* on classification). It will be shown that a 'consistent' coding pattern does not necessarily mean that agreement or consensus between users has been demonstrated.

This is important because agreement (i.e. *consensus* not *consistency*) is a pre-requisite for the pragmatic usefulness of a coding device. Indeed, researchers and practitioners in the area have often alluded to this fact (e.g. Groeneweg 1996: 134; Stanton and Stevenage 1998: 1,746). Put simply, if two people using an HRA technique, Root Cause Coding System or any safety management taxonomy *cannot agree independently* on how to classify individual cases (accidents, errors, incidents) then the technique cannot be classed as 'reliable'. It cannot therefore be useful for classifying features of an event or for targeting remedial action.

Viewing coding schemes as 'reliable' because they generate *consistent* output leads to the possibility that 'unreliable' coding (i.e. disagreement on individual cases) can be overlooked. For example, we have come to the conclusion that terminology may have to be revised so that clarity can be restored. In addition, description of trials on coding systems needs to be more precise, in order that these trials can be evaluated properly.

A discussion of the use of statistical measures of categorical agreement is included at the end of the chapter, and an approach that avoids unwarranted assumptions about 'chance' agreement is recommended. Finally, an outline of the steps necessary to validate coding taxonomies is proposed.

Definitions of reliability

Reliability in relation to consistency and objective standards

In safety management, it has commonly been assumed that there is an external, objective standard to which 'codes' in a taxonomy are supposed to correspond. This 'standard' may be, for example, the codes an 'expert' user would apply, or data from another source, such as an independent investigation. Within this paradigm, the *reliability* of the taxonomy may be defined as the extent to which the codes generated 'agree' with the standard. If disagreement between users occurs, rather than being seen as an indictment of the system, comparison with the objective standard can determine who is 'right' and who is 'wrong'. It is probable that a lack of emphasis on *agreement* in event classification arises from the view that subjective judgments can be scored as 'true' or 'false' in relation to a definitive account.

This definition ('reliability' as measured against some 'objective' source) is adopted from the engineering domain. As relating to a single technical item, reliability simply equates to *1 – the probability of failure*. The Military Standard (1980) definition is 'the duration or probability of failure-free performance under stated conditions' (thus the reliability of an item which breaks down in 1 out of 10 trials is 0.9). Wickens and Hollands (2000: 498–9) describe how reliability in HRA has been seen as analogous to technical reliability (i.e. human reliability is conceived as *1 – the probability of human error*). This formula is explained in more detail by Timpe (1993: 123).

Current 'second generation' devices rely on a similar approach. Hollnagel (1998) states that 'The standard definition of reliability is the probability that a person will perform to the requirements of the task for a specific period of time' (16).[4] So the human 'reliability' in HRA is determined by the extent to which the operator deviates from the required state. Thus reliability is tested (or predicted) by comparing data on possible error situations to data on actual error situations.[5] This engineering definition equates reliability with consistency. Importantly, this means that, as well as *people* being classed as 'reliable' when they are 'consistent', *techniques and taxonomies which produce consistent output are sometimes said to be 'reliable'.*

This appears logical enough, but in fact it leads to a logical confusion. If we call a consistent pattern of 'codes' from a coding scheme *reliable*, there is a tendency to assume that *consensus* has been established, i.e. that people can agree on which codes to pick for individual events. For example, in the context of testing a taxonomy, Stanton and Stevenage (1998: 1,740) define 'consistency' thus: 'this criterion is the same as inter-rater reliability, i.e. the degree to which two analysts make the same error predictions'.

But this equation of consistency with inter-rater reliability is misleading. Coding schemes can actually produce highly *consistent* data in the absence of independent *agreement* on discrete events. Indeed, in the most extreme case, consistent or reliable coding can be demonstrated in the absence of any agreement at all. This distinction is discussed in detail by James *et al.* (1993: 306). The following example shows how this can be the case.

Replicable patterns and unreliable coding

We were recently called upon to work with a 'root cause' event coding system that has been utilised for some years by a section of the nuclear industry. A stated purpose of the system is to codify events to enable identification of trends and patterns within accumulated event data. As part of the work (see Wallace *et al.* 2002), a time based analysis was conducted on the frequency of code assignation over a 22-month period (June 1996 to April 1998). A selection of events (n = 376) from five plants was offered for examination, and a database showing the total cumulative frequency of assignment of particular codes to safety events over the period was created.

The actual codes chosen followed traditional lines (human errors, procedural failures, technical faults, etc.) However, the remarkable feature of the database was that a very similar pattern emerged *regardless of the time period selected*. In other words, the total cumulative code frequencies resembled any number of patterns relating to shorter periods of time within the stated period. So beyond a certain minimum time period the data distribution appeared to be fractal. In addition, further data obtained from three different plants within the industry portrayed a similar picture. Thus distribution appeared to be largely unaffected by time, and the total distribution from five plants resembled the distribution of independent data obtained from three different plants. This distribution is shown in Figure 8.1.

In the course of informal discussion with plant feedback engineers, this consistent output was generally presented as evidence that *the system was reliable*. These codes appear all the time, for all coders, at all plants, therefore they must be 'true'. Note the use of the term *reliability*, meaning the consistent repeated patterns that emerge from use of the system. It was, however, also clear that people thought that the patterns emerging from use

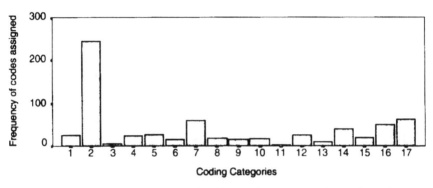

NB Whilst it is the coding pattern that is of interest here, category 2 = work practices, category 7 = procedures and documents and categories 14–17 contain technical codes.

Figure 8.1   The frequency of individual codes (n = 674) assigned to 376 randomly selected event reports over a 22-month period (June 1996–April 1998).

of the system meant *there was consensus* on which codes to apply (as in the 'human science' definition of reliability). The question is, does it follow that the emergence of consistent patterns (i.e. people assign the same codes overall) means that people have agreed on which codes to assign to specific individual events?

The trial of consensus/agreement

Twenty-eight previously coded events were randomly selected from the existing files. Three experienced coders from within the industry were asked to read each event report, additional memos and reports, and to assign causal codes (196 possible choices) in the usual way. The complete data from this study are reported in Wallace *et al.* (2002: 2), however the pertinent point here is that agreement between the engineers as to the 'root causes' of events was relatively low (index of concordance = 42 per cent). The coders clearly did not choose the same codes to a level that satisfies the normal criteria (Borg and Gall 1989). So the answer to the above question is no. Codes can be consistent in the absence of agreement or consensus.

Note that the consistency of the *patterns* in the database had been assumed to be evidence for 'reliability' in terms of inter-rater agreement. However, the results of the reliability testing (low consensus on codes applied to individual events) lead us to suspect that the apparent consistency in the database derives from sources of error variance, or demand characteristics, within the coding and category scheme itself. Strong evidence for such a supposition was provided by analysis of the distribution of codes assigned during the trial. These are shown in Figure 8.2.

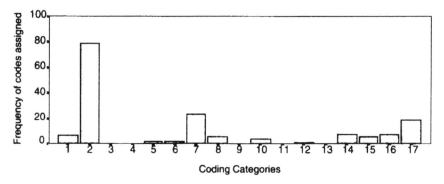

Figure 8.2 The frequency of individual codes assigned during the consensus trial.

The similarity in the *pattern* of codes assigned during the trial to the pattern assigned regularly over random time periods at a number of plants is apparent and striking. A strong correlation was found (rho = 0.837). It remains only to repeat that the second pattern emerged from the *unsatisfactory* consensus trial.

The above data are of general interest because they negate the assumption that replicable patterns (i.e. patterns which *correlate*) are evidence for *consensus* in code or category assignment. As these data clearly show, this is not the case. This suggests, of course, that coding patterns may derive, at least in part, from biases that are properties of a coding system and that conclusions drawn from such a database could be invalid. In addition, any predictive and discriminatory utility of the database is lost. Perhaps the following analogy may be useful in describing how such data come about.

Let us imagine our investigators to be blindfolded and given 50 numbered tennis balls that they each serve randomly at a large board with holes in it. Some holes are bigger than others. On repeated trials of 50 serves for one player, it is observed that on every occasion the bigger holes fill up more rapidly than the others. Similarly, different servers end up with roughly equal numbers of balls in the holes. However, the balls that make up these clusters are never numbered the same (remember the blindfold means the numbers on the balls are random). In this analogy the 'big holes' are the perceptually 'large' categories which attract greater numbers of codes. Thus different people assign events to them differently on different occasions. Nonetheless, because of their 'size' they will always contain more 'hits'. Consensus, however, depends on the *actual codes applied to events* (like the actual numbers on the balls) on specific occasions. Thus it has to be assessed independently. Over the course of the project we obtained evidence from three separate reliability trials that the system did not facilitate reliable coding. This highlights both the misconception that a recurrent pattern of data can only be produced by consensual coding, and the inappropriate use of correlations in testing agreement with schemes involving coding choices (usually called 'nominal scales', e.g. Cohen 1960). The codes assigned in this case (deriving from essentially random processes) could of course be misleading as the basis for courses of action. Assuming that a 'work practice' cause is valid for a given plant event is untenable when reliability is low. Put simply, another investigator might well decide an alternative root cause for the event, requiring a different response. Wallace *et al.* (2002) describe how we worked closely with the industry in attempting to produce an alternative, reliable system on which to base remedial action.

An alternative definition of 'reliability' to that from the engineering/safety management tradition is now described.

Reliability in relation to consensus between subjective sources

Classifying events is a common technique in the human sciences (e.g. psychology, sociology, psychiatry) and has been shown to be highly useful, dependent on certain important criteria. The most critical of these has usually been called 'inter-judge' (Cohen 1960), 'inter-observer' (Caro *et al.* 1979) or 'inter-rater' (Posner *et al.* 1990) *reliability*. Tests of this criterion have usually been simply called *reliability studies* (Grove *et al.* 1981). (The related concept of 'intra-rater' reliability refers to a comparison between the judgments made by the same judge about the same data *on different occasions*.) In the human sciences, 'reliability' equates to the extent to which independent users of a coding scheme, taxonomy or similar diagnostic technique can *agree* on discrete events to be coded (e.g. Caro *et al.* 1979). There are two assumptions implicit here. One is that agreement is to be sought on *individual units* or events, and the second is that both raters are *fallible* (e.g. Spitznagel and Helzer 1985). This is why we have termed 'reliability trials' as *tests of consensus*.

We have shown that confusion can arise during the testing of safety management systems due to this first principle – that agreement refers to the ability to discriminate *for individual subjects, events or cases* (e.g. Cohen 1960; Fleiss 1971; James *et al.* 1993) – being overlooked. If 'reliability' of a coding frame is equated to *consistency in overall patterns*, then misconceptions can arise, where it is assumed without justification that people can apply the same categories to individual events.

The approaches to safety management described in Chapters six and seven rely on consensual accounts of events as the basis for interpretation and action. If we accept the second principle, that agreement between subjective decision-makers is the basic model of scientific interpretation (see Chapter two), then clear evidence that consensus can be reached becomes a vital part of event classification. We view consensus between people (safety managers, investigators and academics) involved in classifying events as the primary building block for the pragmatic usefulness of coding systems and the scientific analysis of safety data.

Inter-rater consensus

In order to suggest a way out of this confusion, we may need to put some 'clear blue water' between studies on consensus between people on the one hand and 'reliability trials' (that do not measure the capacity of users to agree independently on classification) on the other. A start would be to use the term 'inter-rater consensus' (IRC) to denote *agreement between independent users of a technique in relation to classification of individual events* (that is the definition of 'reliability' in the 'human sciences' sense). If adopted, use of such a term can distinguish between techniques where this has been established and those where it has not. (We would see 'inter-rater agreement' as being an equivalent term, however IRC is preferred as an acronym to IRA!)[6]

Let us now describe some of the problems that arise from conflicting definitions of 'reliability', highlighting a lack of clear *consensus* data with which we might begin to validate coding taxonomies.

Problem areas in testing consensus

A basic lack of consensus data

Perhaps the most fundamental problem in this area is that data on inter-rater consensus trials seem hard to come by. It should be noted that other authors have come to a similar conclusion. Wagenaar and van der Schrier (1997) evaluated a number of techniques – including MORT (Johnson, 1980); STEP (Hendrick and Benner 1987); and FTA (Ferry 1988) – and describe these techniques as presenting no 'inter-rater reliability' data. Further they state that 'there is no real excuse for the lack of reliability testing ... since it is not difficult to measure between-raters reliability...' (31).[7] It may be that, in viewing one source as 'objectively true', that designers may have overlooked the need for consensus (agreement between independent users) altogether, and have simply attempted to *validate* each source separately.

Using patterns or frequencies as a test of consensus

We described above how patterns of codes assigned are inappropriate as the basis for which to calculate agreement or consensus. The usual method for calculating inter-rater consensus is to calculate the 'index of concordance', (e.g. Martin and Bateson 1993: 120). This is done by applying the formula $A/A + D$, where A = the total number of agreements and D = the total number of disagreements. Inter-rater consensus can then be reported as a figure between zero and one or as a percentage by multiplying by 100, and is often 'corrected' using coefficients described later in this chapter.

Another common method for calculating consensus is to use a correlation coefficient such as Spearman's rank order correlation coefficient.

Importantly, correlation coefficients for measuring agreement are 'applicable to total frequencies or durations of [...] categories ... but not to individual occurrences' (Caro *et al.* 1979: 306). This assertion can be used as a rough guide to the aspects of a coding scheme that have been tested during a 'reliability trial'. *A reported correlation in the case of a coding scheme where events have to be categorised* cannot be used as evidence that *inter-rater consensus can be achieved*. Instead, consistency of total frequencies of codes has been demonstrated (or otherwise).[8]

But, as we have demonstrated, we need to reaffirm the principle that coding schemes must show inter-rater consensus before they can be considered for use. Correlations alone are not adequate for demonstrating agreement (consensus) between coders. These are only valid with *continuous scales* (for example, where people are required to estimate numbers of errors, or score system factors for importance). Here, correlations are appropriate to show whether estimates are 'agreed upon'. For example, estimates across a range of problems in terms of percentages might be ranked for two 'raters' to show how well they can agree (i.e. how much consensus exists). However, where people simply have to assign codes or categories to events, correlations (in this case based on *frequencies of codes assigned* to individual categories) give an indication of *maximum* agreement only for individual cases.

For example, suppose you and I each assign five codes to a set of events exactly the same number of times overall (e.g. the frequencies in each 'box' are A = 50, B = 40, C = 30, D = 20, and E = 10 for each coder). The rank order correlation will equal one. We are in perfect 'agreement' overall. However, in testing the discriminatory power of a set of categories this is not really what we wish to measure, which is whether we can assign the same code (or codes) to a single event. Within the 'perfect' overall agreement, we might disagree (i.e. assign a different code to) on any number of individual events.

Further examples from the literature

Further examples can be identified where overall frequencies have been used to test 'reliability', leading to inter-rater consensus being overlooked.

Kirwin (1988) undertook a review of five Human Reliability Assessment (HRA) techniques available at the time, which have come to be known as the 'first generation' of such techniques (e.g. Hollnagel 1998). Techniques tested included the Technique for Human Error Rate Prediction (THERP; Swain and Guttman 1983), the Human Error Assessment and Reduction Technique (HEART; Williams 1986) and the Success Likelihood Index Method using Multi-Attribute Utility Decomposition (SLIM-MAUD; Embrey *et al.* 1984).

Kirwin (1988) was concerned that 'two or more independent trials of the technique against the same data would yield approximately the same results' (89). This seemed to imply that he wished to compare raters' scores or codes using each technique *against identical event data*. However, the data presented later in the paper (Kirwin 1988: 99) do not appear to be of this type. 'Consistency' is evaluated by computing correlations measuring *convergence between different techniques.* We have argued that the use of such 'pattern matching' (whether between HRA data and 'industry' data; between data from different taxonomies; between data from different raters) to demonstrate 'reliability' leaves open the question of whether users can *agree* on individual cases, meaning usefulness cannot be established.

Stanton and Stevenage (1998) undertook a major review of the SHERPA technique – Systematic Human Error Reduction and Prediction Approach (Embrey 1986) – in terms of 'acceptability, reliability and validity' and concluded that the technique 'may be acquired with relative ease and can provide reasonable error predictions' (1,737). In the study, comparison in terms of users of the technique was presented in terms of test-retest (or intra-rater) data. 'Test-retest' reliability was used because it 'offered a convenient measure of the consistency of HEI techniques used on separate occasions (Stanton and Stevenage 1998: 1,740).[9] We can immediately see that 'reliability' is equated with 'consistency'; the authors adopt this definition from Kirwin (1992b). Once more, correlations were computed based on frequencies of codes assigned. However, the fact that there is little variance in overall frequencies does not mean people have necessarily rated the same events in the same way the second time around. Caro *et al.* (1979) argue that data calculated using total frequencies are not applicable when assessing the value of coding frames as discriminatory devices, i.e. when we wish to assess whether the categories can be agreed upon. It is not hard to imagine how users could change their minds on individual cases, yet come up with a similar pattern across the four 'boxes' that would produce a similar sensitivity index. So in the example above, if a rater 'hit' a certain error and 'missed' another first time, then missed the first and hit the second next time, the correlation would still show a *consistent* response.[10]

Groeneweg (1996: 229) presents reliability data on the 'TRIPOD' safety management system (see also Wagenaar *et al.* 1994: 2,006). Tables show various rank order correlations for 'System State Profiles' (SSPs) drawn up by different groups of raters (and the same groups on different occasions). The correlations are between mean values (averaged from members of each group) which equate to a 'weighting' of items on a check-list. It is important to note what is being reported here. Users are being asked to generate the SSPs by 'weighting' items (relating to General Failure Types – GFTs; Reason 1990) in terms of relative importance. Hence Groeneweg has used the correct test (correlations are appropriate where continuous rating scales are involved), and has demonstrated high levels of agreement.

So TRIPOD can indeed be shown to be reliable in terms of agreement on *ratings given to pre-determined categories.* However, there is another aspect to the TRIPOD system. It is also intended to allow for the identification of GFTs in accident data. As Groeneweg states, 'The task of an accident analyst is to identify contributing factors and to *categorise* them correctly in the right GFT' (203) (*emphasis added*). Now, instead of rating pre-given items we have to decide which ones to pick. But rank order correlation is not appropriate as a test of this type of agreement (i.e. *categorical* agreement where codes are chosen). No data on classifying using GFTs are presented. However, there are data on classification using a system proposed by Feggetter (1982). In this case, *correlations* are once again used to claim a 'reliability index' (Groeneweg 1996: 135). But it is possible that the raters assigned different codes to individual events, yet ended up with the similar overall patterns shown.

TRIPOD has been shown to be reliable when agreement is sought on the weighting of various items presented in questionnaire format. However, the use of the Feggetter factors (or the GFTs) *as a classification system* cannot be assumed to be reliable on the basis of the inter-rater data presented. High correlations between general frequencies of codes do not mean people will agree when asked to assign codes during event analysis.

Consensus and multiple raters

Problems can also arise in calculating consensus agreement indices. It is usually desirable that an average 'inter-rater consensus' score can be calculated when there are more than two coders involved in a trial. However, a measure of consensus should be computed for each pair of coders separately, so that an average can be computed. A brief note is necessary to show why this is the case.

Silvester *et al.* (1999; on 'organisational culture') and Munton *et al.* (1999; on 'organisational attributions') use a coding system for texts originally developed by Stratton *et al.* (1988). The reliability of the system was tested by Stratton *et al.* who state that '... a total of 315 ... statements was extracted and of these, 220 were identified by independent agreement of *at least two of the three raters*' (89) (*emphasis added*). Fliess (1971) outlines clearly how agreement between multiple raters (in this case three) is to be calculated. Raw agreement arises from 'the proportion of agreeing pairs out of all the ... possible pairs of assignments' (379). Let us examine Stratton's data in this regard.

If 220 out of 315 statements were extracted by 'at least two' coders, there are 95 occasions where a statement was identified by one coder only.[11] Let us consider this latter case first. The assignment here looks like this:

CODER X   Y   Z
CODE   Yes No   No

We can see that coders Y and Z have agreed, whilst coders X and Y and coders X and Z have disagreed. Thus the agreement for the 95 statements would be 33 per cent.

In the case where two coders have extracted a statement, the agreement looks like this:

CODER   X   Y   Z
CODE   Yes   Yes   No

We can see that coders X and Y have agreed, whilst coders X and Z and coders Y and Z have disagreed. Thus the agreement for the 220 statements would *still* be 33 per cent. So when Stratton *et al.* (1988) state that on 220 occasions *at least* two coders agreed, we can see that this merely shows that reliability was *at least* 33 per cent in these cases.

In order to establish whether consensus between raters is acceptable in this case, we would need to know *how many times all three coders agreed*. These authors seem to have assumed 'two out of three' people identifying a statement to be adequate. As they state 'we adopted the more *conservative procedure* of using only the ... statements that were agreed by at least two raters ...' (89) (*emphasis added*). However, without information as to three-way agreements, this cannot really be said to be 'conservative'. In the worst case scenario (no three-way agreement) the total agreement is 33 per cent. Even if, for example, half the 220 statements in the 'two or more' group were agreed by *all three* raters, then the total agreement is 56.39 per cent, which still leaves doubt as to the usefulness of the system.[12]

Pre-selecting events

TRACEr stands for the Technique for the Retrospective and Predictive Analysis of Cognitive Errors, and was developed in the U.K. by National Air Traffic Services (Isaac *et al.* 2002; Isaac *et al.* 2000; Shorrock and Kirwin, In Press). Reliability data for TRACEr were presented at the 24th European Association for Aviation conference in Crieff in Scotland in September 2000 (Isaac *et al.* 2000).

The errors to be classified in the TRACEr trial were highlighted in each incident report – so there was no test as to the consensus on *identification of errors* for coding. This aspect was tested in one of our own trials. Wallace *et al.* (2002) report a trial of SECAS (Strathclyde Event Coding and Analysis System) which was developed for use in the nuclear industry. There were two distinct parts to the trial. In the first part, events to be coded were identified and coded *independently* by raters so that agreement could be tested on both selection and coding of events. In the second part, events for coding were first selected from a sample of reports by one coder, who then coded those events. These events were then passed on to a second coder who coded them independently. In this way, agreement (or disagreement) due to the coding scheme could be tested independently from agreement (or disagreement) as to what constituted a 'codeable' event. Agreement was around 19.5 per cent higher when the second coder did not have to read whole reports and *decide what events to code*. This is important, as most trials reviewed here used pre-selected events. We would recommend factoring in a drop in agreement of at least 10 per cent when users have to identify events themselves, although it can be conceded that, in practice, events to be coded are often decided upon prior to any independent use of a system.[13]

Ambiguity in reporting

Stanton and Stevenage (1998) make reference to a previous work where 'informal data ... suggest a high degree of reliability of SHERPA use between analysts' (1,746). Reliability of between 90 and 100 per cent is claimed in this previous paper (Baber and Stanton 1996: 127).

However, the data presented in the paper (126) are somewhat ambiguous. The system involves identifying errors and classifying them using a coding system for error types. First, it is stated that 'Analyst two found 47 errors, 44 of which were found by analyst one'. This appears at first glance to be evidence for consensual coding (i.e. agreement on individual cases). Then the authors state that 'surprisingly enough, the analysts actually produced the same set of error types, despite differences in the overall number of errors'.

If coders chose *exactly the same types* in all but three cases, (i.e. there was consensual coding in 44 cases), then it is not clear why this would be 'surprising'. This would only be 'surprising' if the first statement simply means that they identified 47 and 44 errors respectively. Then they presented a similar set 'or *pattern*' of error types overall. However, we have shown that this does not equate to 'reliability' as we have defined it (agreement or consensus on individual classifications). Whether this is the case or not, this example shows why clear presentation of results is essential.

Low agreement/consensus

A final issue with tests for consensus in the use of taxonomies arises where the tests and calculations are appropriate and unambiguous but do not lead to conclusive results. In the TRACEr trial, subjects' classifications were compared with 'those intended by the expert developers'.[14] Interestingly, the raw agreement was typically high in relation to the Kappa values obtained.[15] For example, in the classification of 'Cognitive Domain', raw agreement was 87 per cent whilst a Kappa value of 0.46 was obtained. (The relatively small number of choices means a high proportion of agreement would be expected by chance, and the correction is large.)[16] Such Kappa values do not lend themselves to unequivocal support for a system.

Raw agreement on some categories – for example the PSFs (Reason 1990) – actually proved less successful. Isaac *et al.* (2000) call this an 'area of concern', and undertake to work towards achieving higher consistency in further development. Taken with the low Kappa values presented, the trial cannot be said to be entirely satisfactory. Nevertheless, we would welcome the developers making the trial public, and for reporting agreement in different areas of the scheme rather than giving an overall figure which can serve to cover up important aspects of disagreement.

Finally, over the course of our work with CIRAS and SECAS[17] and in developing classificatory systems in the field of occupational and health psychology (Davies 1997), we have come to be concerned with the common use of the Kappa coefficient (Cohen 1960, 1968; Fleiss 1971). This is typically used in the human sciences to correct the raw proportion of agreement (Index of Concordance) for agreement 'expected by chance'.[18] A discussion seems wise as to the applicability of this technique for calculating the reliability of safety management schemes.

Statistical measurements of inter-rater consensus

Cohen's Kappa

The Kappa coefficient (Cohen 1960) has been widely used, principally in the medical field after a review of inconsistency in clinical methods by Koran (1975). In keeping with the distinction outlined above between correlation and consensus on individual cases, Kappa is used precisely because it can be interpreted 'as a measure of the amount of *agreement (as opposed to correlation or association)* between two raters ...' (Spitznagel and Helzer 1985 *emphasis added*). However, Kappa has been extensively critiqued, and its use in this context is questioned here.

Kappa is a simple formula for correcting the number of categorical agreements between independent judges for the number of agreements that would be expected purely by chance. The probability of chance agreement on a single code is calculated from the probability of each rater using a code relative to the total number of codes assigned (Cohen 1960: 38). Kappa is then computed as:

$$[(A/A + D) - \text{chance agreement}]/(1 - \text{chance agreement})$$

where A = number of agreements and D = number of disagreements.

Problems with Kappa include fundamental issues with the assumptions behind the statistic and practical problems including those of misinterpretation and misuse.

Kappa and chance agreement

It is useful to note that Cohen (1960: 38) is quite clear as to the conditions under which Kappa can be used.

1   The events/scenarios/errors to be coded are independent.
2   The categories used are independent, mutually exclusive and exhaustive.
3   The coders operate independently.

Perhaps the most fundamental criticism of Kappa is that the concept of correcting for 'chance agreement' is inherently flawed (Grove *et al.* 1981; Carey and Gottesman 1978; Janes 1979). This is related to violation of the third condition above, independence of raters. Kappa can be used in two ways. First, it can be used to test *independence* of coders (i.e. as a test statistic). This involves testing the data against the (somewhat unlikely) null hypothesis that coders' decisions will be completely unrelated to each other.[19] Secondly, and more commonly, Kappa is used as a way to quantify the *level of agreement* (i.e. as an effect-size measure). This second use is based on the proportion of chance (or expected) agreement. However, the term 'chance' is relevant only under the conditions of *statistical independence* of raters. If coders are not independent then Kappa cannot be used to correct raw agreement. So, first, Kappa must be used to test whether Kappa can be used!

Usually, the first use would determine that coders (and therefore the decisions that they make) are *not* independent. Thus, the common statement that Kappa is a 'chance-corrected *measure* of agreement' is misleading. As a test statistic, Kappa can verify that agreement exceeds chance levels. But as a *measure* of the level of agreement, it is not clear that Kappa 'corrects for chance'. As Spitznagel and Helzer (1985) put it 'the assumptions about the independence of errors [i.e. coders' decisions] is probably never correct' (727). Maxwell (1977) argues that it is absurd to argue that coders start from a position of complete ignorance, and thus rejects the idea that Kappa can measure agreement relative to the proportion of cases arising from chance alone.

Kappa and mutually exclusive codes

Another assumption underlying the use of the Kappa statistic is that codes tested will be mutually exclusive and exhaustive codes. There appears to be a paradox here. One essentially tests how much overlap and redundancy there is in a coding taxonomy by calculating agreement. If trained coders cannot agree on classification then we would argue that the codes are not functioning as exclusive categories. Yet mutually exclusive categories are a prerequisite for Kappa. There is no easy solution here – developers must simply endeavour to design coding frames where definitions are clear and choices are absolute. As a guideline here, we would recommend computing *raw agreement* – i.e. the Index of Concordance, (Martin and Bateson 1993) – first. Kappa coefficients calculated on the basis of *low raw agreement* can be assumed to violate the principle of mutual exclusivity and should be avoided.

Kappa and prevalence

Another problem with Kappa is that it is sensitive to prevalence in codes assigned, irrespective of agreement. An example can be shown using the dichotomous yes/no classifications for which Kappa (Cohen 1960) was originally designed. It is also important to draw up contingency tables which include an aspect that is often overlooked, the cases where we *agree* on a code by *leaving it out*.

Let us imagine a basic trial involving two coders, you and me. If you assign code 'A' given five choices A, B, C, D and E it means you have said 'yes' to 'A' and 'no' to B, C, D and E. So if we agree on the code for this single case (i.e. *I say A as well*), then we have five agreements (one 'yes-yes' and four 'no-no'). If we disagree on the *assigned codes* (e.g. *you say A and I say B*) then we have two disagreements (yes-no for A and no-yes for B) and three agreements (no-no for C, D and E). This of course has implications for raw agreement, which will be a function of the number of codes available. Over a series of attempts, raw agreement will be high for codes we don't use much (for example, 100 per cent for a code neither pick at all). However, it can be argued that agreement is best calculated for individual codes anyway, so sources of disagreement are not lost in an aggregate (Vitez *et al.* 1984). In this way, the *qualitative* aspects to agreement will emerge. These aspects (for example to know we can never agree on a code when one of us assigns it but we can always agree to leave it out!) are important to know about. So including the 'no-no' cases (i.e. by drawing up two by two contingency tables for each code and coder – see Tables 8.1 to 8.3 below) may be the best method. Most statistical packages allow for quick and easy production of such contingency tables.[20]

The problem of Kappa's sensitivity to prevalence (also called base rate; Spitznagel and Helzer 1985), has been extensively discussed in the literature on psychiatric diagnosis and epidemiology. In simple terms, the problem is that Kappa varies not just with consensus between raters but with distribution of cases to be coded. Grove *et al.* (1981) point out that Kappa is actually a series of reliabilities, one for each base rate. As a result, Kappas are seldom comparable across studies, procedures, or populations (Thompson and Walter 1988; Feinstein and Cicchetti 1990).[21]

To illustrate the prevalence problem, we can assume that in any given population of safety events some causes will be attributed more than others.[22] Let us then state that we have different Kappa values (chance-corrected measures of agreement between investigators) for different 'codes'. Suppose the Kappa value for code A is twice as high as the value for codes B and C. In such cases, it is flawed to assume that we *necessarily* agree better on code A. This is because the *K* values obtained are highly sensitive to the *prevalence* of each type across the events in question. Kappa will rise as prevalence reaches an optimal level and then decrease again as prevalence increases beyond this. Thus particularly low or high prevalence lead to lower Kappa values *irrespective of agreement*. We have provided a worked example to show this effect. We are indebted to Feinstein and Cicchetti (1990: 544–5), as we have essentially adapted the example they provide to put it in a safety management context, without going into as much detail with the statistics.

Suppose two investigators rely on a taxonomy containing *'human error'*, *technical* and *environmental* codes. These investigators decide to study 100 event reports to determine whether these causes were present. Appropriately, they decide to measure how well they agree on the causes of previous

Table 8.1 Assignment of technical causes to 100 events by two investigators

		Investigator A Yes	No	Totals
Investigator B	Yes	40	9	49
	No	6	45	51
	Totals	46	54	100

events, and to use Kappa so as to be certain their classifications are higher than expected by chance. Table 8.1 shows their assignments (present or absent) concerning the *technical* causes.

We can see that the investigators agreed that 40 crashes were due to technical factors and 45 were not (this is the no-no agreement outlined above). They disagreed on fifteen reports, with six identified as technical failure by investigator A and not B, and nine assigned by B and not A. The raw agreement Po = 85/100 = 0.85. The marginal distribution leads to a Pe (chance agreement) of $(0.46)(0.49) + (0.54)(0.51) = 0.5$. Thus Kappa becomes:

$$K = (Po - Pe) / (1 - Pe) = (0.85 - 0.5) / (1 - 0.5) \quad = 0.35/0.5 = 0.7$$

So we have a raw agreement of 85 per cent and a Kappa coefficient of 0.7.

Now let us look at their agreement for the *environmental* factors in Table 8.2. We can see that the investigators agreed that five crashes were due to environmental factors and 80 were not. They disagreed on fifteen reports, with five identified as containing environmental factors by investigator A and not B, and ten assigned by B and not A. The raw agreement Po = 85/100 = 0.85. The marginal distribution leads to a Pe (chance agreement) of $(0.1)(0.15) + (0.9)(0.85) = 0.78$. Thus Kappa becomes:

$$K = (Po - Pe)/(1 - Pe) = (0.85 - 0.78)/(1 - 0.78) = 0.07/0.22 = 0.318$$

So we have a raw agreement of 85 per cent and a Kappa coefficient of 0.318.

Finally, let us look at agreement for 'human error' causes shown in Table 8.3. We can see that the investigators agreed that 80 crashes were due to human error and 5 were not. They disagreed on 15 reports, with 5 identified as containing human error by investigator A and not B, and 10 assigned

Table 8.2 Assignment of environmental causes to 100 events by two investigators

		Investigator A Yes	No	Totals
Investigator B	Yes	5	10	15
	No	5	80	85
	Totals	10	90	100

Table 8.3 Assignment of 'human error' causes to 100 events by two investigators

		Investigator A Yes	No	Totals
Investigator B	Yes	80	10	90
	No	5	5	10
	Totals	85	15	100

by B and not A. The raw agreement Po = 85/100 = 0.85. The marginal distribution leads to a Pe (chance agreement) of (0.1)(0.15) + (0.9)(0.85) = 0.78. Thus Kappa becomes:

$$K = (Po - Pe)/(1 - Pe) = (0.85 - 0.78)/(1 - 0.78) = 0.07/0.22 = 0.318$$

So we have a raw agreement of 85 per cent and a Kappa coefficient of 0.318. Note what has happened here. In each of the three cases *raw agreement is the same*, and we have the same number of 'codes' to choose from (yes and no). However, in the human error and environmental cases, we have a lower Kappa value than for the *technical* factors. The probability of agreement by chance is deemed to be higher in the former cases because each investigator is more likely to assign these to a particular 'box' (i.e. present or absent) than is the case with technical factors. So *prevalence* is shown to affect Kappa irrespective of *consensus*, which is what we are interested in.

It is clear that we cannot be satisfied with lower Kappa coefficients for human or environmental factors just because our investigators agree there are a lot of them, (or not very many of them) in our reports.

Alternatives to Kappa

Alternatives to Kappa have been proposed. A solution to the prevalence problem was proposed by Spitznagel and Helzer (1985), who also provide a detailed discussion of these issues (see also Cicchetti and Feinstein 1990). Spitznagel and Helzer recommended a 90-year-old statistic called the coefficient of colligation (Y) (Yule 1912). Y remains more stable than Kappa (it is effectively independent of prevalence) for all but high prevalence rates. Suffice to say that we have not seen Y values computed for agreement in safety management work, probably because the formula is more complicated than Kappa. The Y coefficient can be calculated at 0.7 for Table 8.1, and 0.48 for Table 8.2. So whilst it seems to be more robust than Kappa, it suffers from the same problem in that high and low prevalence affects the coefficient (Spitznagel and Helzer 1985).[23][24]

Lee and Del Fabbro (2002) argue that a fundamental problem with Kappa and Y is that they are based on a 'frequentist' approach to probability estimation. So probabilities are equated with frequencies, and 'available

$$BK = 2 \left\{ \frac{a+1}{a+b+2} \quad \frac{a+b+1}{n+2} + \frac{d+1}{c+d+2} \quad \frac{c+d+1}{n+2} \right\} - 1$$

(Where a, b, c, and d are cell totals in a contingency table
(see Tables 8.1, 8.2 and 8.3 above) and n = a + b + c + d).

Figure 8.3 BK coefficient of agreement for binary decisions.

counts from a finite data set are used to form ratios as estimates of the underlying probabilities' (10). So, 'ten agreements and ten disagreements' is treated the same as 'fifty agreements and fifty disagreements' in the calculation. If a Bayesian approach is adopted (Carlin and Louis 2000; Leonard and Hsu 1999; Sivia 1996), the observed data are used to revise prior beliefs. So, as a picture of agreement builds up, changes in *frequencies* rather than *ratios* alter the calculation (i.e. the more *actual* agreements there are the higher the coefficient becomes). Using a standard result, (Gelman *et al.* 1995: 31), Lee and Del Fabbro derive the 'BK' coefficient of agreement for binary decisions. The formula is presented in Figure 8.3 for reference. It can be seen that it is fairly simple to compute.

The BK coefficient of agreement for Tables 8.1, 8.2 and 8.3 comes out *the same* at 0.67. This is because BK is not sensitive to prevalence, and for the same agreement the coefficient is stable irrespective of marginal totals. BK is, however, sensitive to sample size, and so will become less conservative for the same agreement over larger trials. Moreover, there is no problematic assumption about random assignment by chance against which to compare the patterns. BK is also fairly easy to compute. Whilst recognising that BK is still under review, we would argue that it has been useful in our calculations of agreement to date and we would recommend consideration of its use. (Lee and Del Fabbro are happy to be contacted in this regard at the Department of Psychology, University of Adelaide.)

Finally, let us outline steps for the testing of taxonomies for inter-rater and within-rater consensus.

Procedures for establishing inter-rater consensus (IRC) and within-rater consensus (WRC)

If we accept that inter-rater consensus (IRC) is a requirement for a valid technique, we can prescribe the steps needed in consensus testing and reporting. Our suggestion as to these steps is shown in Table 8.4.

Once these steps have been taken, if it is the conclusion that adequate consensus has been demonstrated in the use of the codes, then tests of the practical usefulness of the system as a diagnostic tool or basis for action might begin at that point. However, any such pragmatic tests prior to such a demonstration could produce highly misleading results.

Table 8.4 Steps for validating coding categories as a safety management tool

Testing for consensus	
Step	*Reporting*
1 Test IRC, the extent to which raters can agree independently on classification of individual events. This should first be calculated as an index of concordance (see above) for each pair of coders.	Make clear the conditions under which the test took place, number of coders, number of events, discussions which took place before and after.
2 Agreements in different areas of the scheme (i.e. for individual codes) should be calculated separately.	If an average agreement is reported, describe in qualitative terms agreement for areas of the scheme, or individual codes and coders.
3 If the index of concordance is satisfactory(see Borg and Gall 1989), calculate the reliability coefficient.* If not, redesign the scheme and start again from step 1).	Report coefficient. If an average is given, describe range for different coding decisions.
4 After a suitable time, test WRC in the same way as steps 1 and 2.	Report as steps 1 and 2.

* BK has been recommended (Lee and Del Fabbro 2002), and we are currently revisiting some of our raw agreement data to calculate BK values.

Summary

If we take on board arguments presented elsewhere in the book about functional discourse, subjectivity and attribution we can see the importance of agreement between 'subjective' decision makers using coding schemes to classify events in safety management. The concept of 'inter-rater agreement' comes from areas (i.e. diagnostics) where there is assumed to be no 'definitive' judgment against which to compare results. As others have recognised, 'HRA' type taxonomies are similarly constrained in that they rely in principal on subjective judgments (Groeneweg 1996).

We have shown how consensus between users of taxonomies may be overlooked (see reviews by Kirwin 1992b; Wagenaar and van der Schrier 1997). It may also be examined inappropriately by looking at correlations between frequencies of codes assigned (Stanton and Stevenage 1998; Groeneweg 1992). We have presented data (Wallace *et al.* 2002) to demonstrate why this does not provide evidence for consensus, and proposed a term [inter-rater consensus (IRC)] which may avoid current confusion around the term 'reliability'.

An examination of the use of statistical measures of agreement has led us to recommend a new Bayesian statistic called BK (Lee and Del Fabbro 2002) which avoids problems with, for example, Cohen's Kappa coefficient (Cohen 1960).

Finally, we have outlined steps we would recommend in testing coding devices, taxonomies and error prediction techniques, emphasising that consensus is a prerequisite for usefulness. Chapter six shows an example of a trial where, we hope, it is clear exactly what has been tested and how.

9 Error taxonomies and 'cognitivism'

One of the most important aspects of a safety manager's task is to classify and log all the situations that are generally described under the rubric of 'human error'. Consequently, a great deal of work has been done on error taxonomies. The most widely cited system in this respect is probably the taxonomy developed by Jens Rasmussen in the late 1970s, the Skill Rule Knowledge (SRK) distinction. This chapter will discuss the origins of this taxonomy, its usefulness, and the question of whether the approach it uses is still the best one for safety management.

The first question to ask, therefore, is, 'what *is* the SRK distinction?' Unfortunately this isn't as easy as it might sound, as the interpretation of SRK given in many safety management texts bears little relation to the taxonomy that Rasmussen actually proposed. For example, it is frequently assumed in the literature that a 'rule-based' error refers to the rules (i.e. formal procedures) of the organisation, that a 'skill-based' error involves someone having insufficient skill for the task, and that a 'knowledge-based' error is an error where the operator did not have sufficient task knowledge. However, this is a misrepresentation. Rasmussen's taxonomy is a tightly defined model based on that branch of information processing theory generally termed 'cognitivism', and an analysis of his definitions of terms shows that the correct definitions of skill-based, rule-based and knowledge-based behaviour frequently bear little or no relation to the way the taxonomy is usually used.

Be that as it may, the key questions to be asked are, what is SRK, how was it developed, and is it still the most effective approach 30 years on? This chapter will attempt to answer this question by looking at the intellectual roots of SRK and by considering the empirical evidence for its validity. However, as should become apparent, a critical discussion of the SRK framework has broader implications for the 'cognitivist' model of cognition widely used within the safety field. The deeper question that underlies this chapter, therefore, is: 'Are "cognitivist" models the best currently available for describing operator behaviour?'

Origins

Rasmussen's early work was based on Verbal Protocol Analysis (VPA) of the discourse of operators in the nuclear industry in the early 1970s. It was in the late 1970s and early 1980s that he first propounded his now classic 'Skill Rule Knowledge' (SRK) taxonomy. This was at the high point of the 'Cognitivist' phase of psychology (Staddon 1999), and its 'cousin', traditional Artificial Intelligence (now referred to, somewhat ironically, as Good Old Fashioned Artificial Intelligence (GOFAI) (Dreyfus 1994)). Since then, Rasmussen has expanded on his basic theory and added to it concepts taken from ecological psychology (specifically the work of J. J. Gibson) and even connectionism (i.e. neural nets/parallel processing). However, despite these additions, the intellectual core of Rasmussen's model (that is, the fundamental SRK distinction) was taken directly from cognitivist models of the functioning of the human brain which were commonly used in the 1960s and 1970s, specifically work done by Newell and Simon (1972) on problem solving and Fitts and Posner (1967) on skill acquisition.

Cognitivism

The basic thrust of Rasmussen's early work is simple. Rasmussen attempts to create a series of distinctions (a taxonomy) which will enable man-interface action (and therefore man-interface error) to be classified. In order to do this he makes two main assumptions.

The first (and most important) assumption Rasmussen makes is that human beings are internal (or cognitive) rule-following organisms: 'the sequential, conscious data processes ... process rules ... are therefore necessary to activate and control the steps in a sequential data process' (Rasmussen 1980: 74). This is posited as an *a priori*, and it should be noted that it is a very old belief: it was originally stated by Plato, and has its roots deep in Western thought (Horgan and Tienson 1996).

At this point two different meanings of the word 'rule' must be distinguished. No one doubts that human beings follow rules: in fact, our social world is constituted of them, and it is difficult to see how society could function without rules of human conduct. However, for Rasmussen, and others in what Lakoff and Johnson (1999) call the 'first wave' of cognitivist psychology, as well as these social rules there are also *internal* rules, usually conceptualised as being similar to the series of instructions in a computer program: that is, as algorithms or flowcharts (Rasmussen 1987b: 54). Rasmussen makes clear that the rule may *originate* externally (that is, from instructions or training), but it is stored in memory, and that, therefore, it is accessed via internal 'cognitive processes' from an internal memory store.

Secondly, Rasmussen's theory of cognition hypothesises the existence of *mental models*. It should be pointed out that the specific theory of mental models used here is descended from a broader theory of 'cognitivist representationalism'. This is the hypothesis that humans use *inner representations* in order to engage in any mental activity. Rasmussen again treats this as an *a priori* assumption: 'Purposive human behaviour *must* be based on an internal representation' (Rasmussen 1983: 258) (*emphasis added*), and slightly earlier he makes clear that by representation he means a mental model (see Rasmussen 1979 for precise elucidation of his concept of 'mental models').

The thrust of these early papers of Rasmussen therefore is to bring these two theories together and hypothesise human beings as *rule-following organisms making use of mental models to solve problems and perform tasks* as the basis of his error classification. This view of cognition is termed 'cognitivism' (Varela *et al.* 1992). In this model, the brain is similar to a digital computer, processing data (which are stored in binary form as 'ones' or 'zeros': 'bits' of information) sequentially via algorithmic rules. The information thus processed is then stored in the form of symbols, which are of two kinds. First there are internal symbols ('states', that is attitudes, opinions and memories are claimed to be of this sort). Secondly there are symbols of external reality ('representations') of various kinds, one of which (the representation of the various systems the operator has to work with) is what Rasmussen means by a mental model (Lakoff and Johnson 1999).[1]

Rasmussen developed his own particular brand of cognitivism, as he states, from a specific research project, that of Newell and Simon, who had developed a computer program (the GPS or General Problem Solver), to attempt to model, and, therefore, explain, human problem solving techniques (Newell and Simon 1990). Rasmussen's theory is in some respects more complex than Newell and Simon's. They were merely attempting to model human problem solving, whereas he is attempting to create a model of all operator behaviour. However, *in most essential respects* Rasmussen's SRK model was derived from Newell and Simon's. Rasmussen states that AI models (by which he presumably means those of Newell and Simon: no other authors are referenced), are 'the ... best available tool for simulation of human information processing' even though he acknowledges that 'AI models have severe limitations' (Rasmussen 1986: 188). As will be demonstrated, the Newell and Simon model does indeed have 'severe limitations'. The question is: is it really, therefore, the 'best available tool for simulation'?

The Skill Rule Knowledge (SRK) distinction

Rasmussen's theory is a theory about human error, and he begins his argument by painting a picture of a single human operator operating technical equipment (that is, he bases his model on the nuclear operators whose working he had studied – Rasmussen here is influenced by Newell and Simon who explicitly repudiate the 'sociological approach' (Newell and Simon 1972)).

As mentioned previously his early work was done on the VPA of nuclear operators:[2] however, what is striking about these early papers is how little reference he makes to this work. Instead they consist of highly abstract and theoretical work on human cognition. The idea, obviously, is to create a theoretical or even philosophical model that will be valid for *all* human–machine interactions, and to do this he drew on the most up-to-date cognitive models that were available to him at the time.[3]

The taxonomy that Rasmussen suggests is based on *degrees of familiarity with the task*. He argues that there is a tripartite distinction of task performance. Tasks with which the operator is extremely familiar are at the *skill-based level*. Tasks which are familiar (but not automatic) are at the *rule-based level*. And unfamiliar tasks are at the *knowledge-based level*. All three of these are internal cognitive states with their own specific cognitive processes. Skill-based behaviour (SBB) makes use of a dynamic mental model. Rule-based behaviour (RBB) uses algorithmic 'rules'. And knowledge-based behaviour (KBB) uses both rules *and* mental models. Thus all task acquisition begins at the knowledge-based (KB) level, then, when familiarity increases, progresses to the rule-based (RB) level. Finally, operators function at the level of skill-based behaviour (SB) when they know the task 'backwards'.

The main support Rasmussen produces for the SRK distinction apart from his own research is the work of Fitts and Posner (Rasmussen 1983: 259). In the 1960s they suggested a three-stage model of skill acquisition: the cognitive stage (where the subject is new to the task), the associative phase (where subject is learning the task) and the autonomous phase. Now, the Fitts and Posner model, (which was created in the late 1960s at the high point of 'cognitivism') is based (like Newell and Simon's) on the 'brain is a digital computer' metaphor (Fitts and Posner 1967).

Given this, it is important to note that the Fitts and Posner model of learning is not the only one available. For example, Gentile (1972; 2000) has proposed a *two-stage* model. In this model the novice begins by 'getting the idea of the movement': determining relevant and irrelevant stimuli and finding the most appropriate movement pattern. Only after this has been done is there a process of fixing and diversification, adapting the operator to the changing environmental demands of the task. So Gentile's model is explicitly based on a systems ('cybernetic') approach to learning and cognition (Miller *et al.* 1960).

In other words the Gentile model stresses *dynamic situated action* rather than internal cognitive 'rules' as with Rasmussen/Fitts and Posner: moreover, Gentile explicitly repudiates the 'computer' metaphor. There is no space here for a detailed comparison of the two theories.[4] However, the key point is that simply detailing the Fitts and Posner model does not prove that SRK is therefore correct: the Fitts and Posner model is contested.

Skill-based behaviour (SBB)

To begin to analyse SRK, we should look at the various levels in order, beginning with skill-based behaviour. Skill-based behaviour is automated behaviour where the operator is fully conversant with the task. Rasmussen states that the skill-based level uses *only* 'a very flexible and efficient dynamic internal role model' for cognition (Rasmussen 1983: 259): therefore, not rules.[5] It should be noted that, contrary to what is usually stated, Rasmussen does not believe that errors at this level *are* errors. They are purely mismatches between the operator and the environment: the operator is *never* to blame. The ubiquitous references to 'skill-based error' in the literature are therefore all incorrect: there is no such category. (Rasmussen writes: 'skill-based behaviour ... is controlled by physiological laws ... and the concept of error becomes meaningless ... misfits in abnormal situations due to effective adaptation to normal system behaviour cannot reasonably be referred to as operator error' [1980a: 109].)

So how is this mental model used at the level of SBB to be created? A new mental model in a new situation must be 'built' at the level of KBB, because that is the level of behaviour Rasmussen states that is used when the operator faces a 'new situation'. However, he provides no mechanism for direct transmission of a mental model from the level of KBB to SBB. The 'middle' level (RBB) uses only rules and not mental models. So even if the model could be build up, it can never be 'passed down to' the level of SBB.[6] Whether the model could be built up at the level of KBB in the first place will be discussed later.

Rule-based behaviour (RBB)

Rasmussen's theory of a 'rule-based' level of human cognition (rule-based behaviour: RBB) most closely resembles Newell and Simon's theories. Here, 'higher grade' activities take place, of an 'if-then' format. Internal stored algorithms are selected and function in a 'familiar work situation' (Rasmussen 1983: 259).

Processing at this level is digital (using discrete 'bits' of information), and sequential. In other words, activities are broken down into steps and followed one after another. So according to this hypothesis when I (for example) walk across the room, there is a programme in my brain that tells me, step 1, put one foot in front of other. If no obstacle, put other foot forward (step 2). And so on (Rasmussen 1982).

There are two main problems with the 'rule-based behaviour' hypothesis: the problem of the fact that rules can be broken down into more fundamental elements, and the problem of relevance.

To take these points in order. If an operator is following a rule (for example, if signal is at red, stop the train), it is clear that this rule can be broken down into elements (for example, red, signal, train), which need to be defined, or else s/he can't follow the rule. At the level of RBB, this must be done by rules. So to follow the above rule, s/he needs to access other rules that define (for example) a signal as an electrical device used to indicate whether one can move the train on a railway track. But this rule in turn can be broken into component elements (e.g. electrical, move, track) which need further definitions in the form of rules And so we are in a situation of infinite regress.

The other problem is relevance. When should the operator stop following one rule and begin to follow another? Obviously, when incoming information makes such a change necessary, that is, when it is task relevant (so in the above example, the operator stops following the rule 'drive train forward' when s/he sees a signal at red). But, in an open dynamic system (like the workplace) the amount of information that might be task relevant is *literally infinite* (given that there are an infinite amount of possibilities that might lead to a task relevant situation (Toft 1996)). And, unless the operator has accessed the rules to define whether this information *is* relevant to the task (and, therefore, task change) all of it *might* be relevant to the task. So *all* the information must be processed by rules that assess whether it is relevant. And all *these* rules can in turn be broken down into elements as above (for example, a definition of 'relevant'). So operators would need infinitely large memory stores (to store all the rules) and infinitely long periods of time to assess all incoming information (Dreyfus 1994). It is clear that this is an impossible model for operator behaviour and it is, in any case, incompatible with current brain research as to how the brain works (Margolis 1987).[7]

Knowledge-based behaviour (KBB)

Rasmussen's concept of knowledge-based behaviour (KBB) was created to deal with an obvious problem of cognitivism: there are a finite number of rules which can be stored in the human brain (assuming this is a valid model of any form of cognition). Therefore, human beings should only be able to deal with a relatively small area of rule-governed activity. In actuality, however, humans can deal with relatively unfamiliar situations for which no pre-stored programme or rule would seem to be available. The 'knowledge-based' sphere of activity therefore has to be introduced to deal with this problem. The theory is that humans are rule followers until they meet with an unusual situation, in which case they create a new rule from first principles, which is then tested either empirically or theoretically – presumably 'internally' on a mental model, which Rasmussen states is also used at this level (Rasmussen 1987b: 55). Rasmussen stresses that this state is energy intensive, and the organism will default to the RBB or SBB levels as soon as possible: 'This [i.e. KBB] is the level of intelligent problem solving which should be the prominent reason for the presence of human operators ... Behaviour in this domain is activated in response to unfamiliar demands from the system. The structure of the activity is an evaluation of the situation and planning of *a proper sequence of actions* to pursue the goal ...' (Rasmussen 1980a: 110) (*emphasis added*). This is the realm of 'free will', and it is obvious why the 'rule-generating' mechanism is left vague, for how is a rule-bound organism to create a new rule spontaneously?

The answer seems to be that it cannot. Rasmussen later acknowledges this when he writes: 'Mental operations at this [i.e. KBB level] ... must be controlled by a complex set of process rules' (Rasmussen 1979: 37). So it seems that in the final analysis KBB 'rule creation' is as rule bound as RBB and is therefore open to the same objections as were raised to RBB. All cognition is either rule based or based on mental models after all. There is no 'third way', which can 'create rules' without rules or mental models.

Even without these philosophical objections, one may well wish to question the model of KBB and its distinctiveness from SBB. Take a relatively novel and unique experience: learning to drive. The novice driver's foot slips off the clutch and the car stalls. What class of error is this? By the logic of the Rasmussen system this must be a knowledge-based error: that is, an error which uses a 'higher conceptual level, in which performance is goal controlled and *knowledge-based* ... the goal is explicitly formulated, based on an analysis of the environment and the overall aims of the person' (Rasmussen 1983: *emphasis added*). It must be of this sort because whilst attempting an unfamiliar task *all* errors are knowledge-based errors. Does this seem plausible? When one sees a novice driver's foot slipping off the clutch, do we assume that s/he is making an error based on his/her 'overall aims'?

Moreover, knowledge-based thinking is long-term, strategic, goal-orientated, planning-type cognition. It is unclear when one would engage in such a cognitive state. However, one might be reasonably certain that it would not be in a situation when a train driver travelling at 150 m.p.h sees a signal fail in a completely unexpected and novel way, and feels he has to brake as hard and as fast as possible. However, according to SRK, this is precisely when one would engage in knowledge-based thinking: in an unfamiliar situation. This has been referred to as the 'Hamlet' model of human cognition: that there is always unlimited time to make decisions. Its relevance to real world safety situations must be questioned (Falla 1999).

However, it might be argued that although the ideas which lie behind SRK are problematic, it may still work in practice as an error taxonomy. For example it may still be possible reliably to assign errors to the skill, rule or knowledge categories. Alternatively there may be empirical data which show that, although it is impossible that the brain works in the way hypothesised by Rasmussen, considered as a *metaphorical paradigm*, SRK may make correct empirical predictions.

Empirical evidence

In the 1990s, numerous empirical tests took place of Ecological Interface Design (EID), an interface based on the SRK taxonomy. It was empirically tested by Vicente (Vicente *et al.* 1995) as the interface for DURESS (DUal REservoir System Simulation), a simulation of a reservoir water control system. Vicente has described two experiments which were carried out on a version of the EID tailored for use for DURESS. In the first experiment, subjects were grouped into experts and novices (based on their theoretical knowledge of the system), and were shown 'a dynamic, real time event sequence of the behaviour of DURESS' (Vicente *et al.* 1995: 532). Subjects were asked to remember as much of the meaning of what had happened as possible, and were then asked some questions pertaining to what they had seen, especially 'what (if anything) was going wrong?'. They were then graded on diagnostic accuracy. As might be expected, the 'experts' performed better than the 'novices'. However, they did not perform that much better. There was considerable overlap between the two groups.

A second experiment was therefore carried out to investigate this result further, which took more variables into account in terms of predicting diagnostic performance. It should be stressed at this point that the DURESS scenarios were picked specifically to create knowledge-based behaviour (KBB): i.e. behaviour where previous experience and knowledge would be of no use.

Given this, it is startling that in this second experiment '[t]he predictors most strongly correlated with performance were previous knowledge of the DURESS system and the ... interface' (Vicente *et al.* 1995: 540). In fact these were just about the only effective predictors of performance. But this directly contradicts the KBB hypothesis, in that at this level operators were supposed to be working from 'first principles' to *create* new rules: *not* from previous knowledge or experience.

It may be argued that perhaps these operators had a 'mental model' of the system, which helped them to think in a manner consistent with KBB. But this misses the point. At no point did the operators think in the 'correct' step-by-step goal directed (KBB) manner as predicted by Rasmussen's taxonomy. Instead they operated in a fashion more consistent with what Vicente termed 'rule-based behaviour', as Vicente admits (Vicente *et al.* 1995: 540). Moreover, DURESS knowledge and interface experience was highly correlated with performance on fault trials but not on normal trials. Again, this is exactly the opposite of what one might expect. 'Normal' situations should be operated on with RBB, where previous experience is essential. In fact, according to SRK, without knowledge of rules and procedures, performance at the RB level cannot take place. It is at the KB level, where thinking is from first principles, that previous experience is of less importance (for further discussions of these experiments and others see Følstad 1999). SRK would predict that DURESS knowledge should be highly correlated with 'RBB' activity, not 'KBB'.

Moreover, SRK directly contradicts other empirical work on 'expertise' which suggests that expert's abilities are highly domain specific, and predicated on previous experience. Therefore experts *never* use 'first principles' cognitive strategies (such as KB, where the principles are general and could theoretically be used in any situation), but instead simply use their specific, previously acquired knowledge and experience to perform tasks (Gilhooly *et al.* 1988; Green and Gilhooly 1992).

So, contrary to what is claimed by the SRK model, there is abundant evidence (Følstad 1999) that the major predictive component of 'expertise' is *experience*. The more someone performs a task, the better they get at it; in other words, experts *always* use previous experience in terms of analysing situations. They know more about the system. That is what makes them experts. Experts never move into KBB, because they *never* reach situations in which there are 'no know how or rules ... from previous encounters'. And neither does anyone else.

And if KBB cannot exist, then SBB cannot exist either, because KBB is essential to 'create' the mental model which is then 'passed down' to the SBB level. And, as we have seen, the level of RBB is conceptually impossible. Therefore the taxonomy is not adequate as a model of operator function.

A final point should be made. It has been demonstrated that SRK does not function as an adequate taxonomy of man-machine interaction. However, will it still function as an error taxonomy? That is, can people use it accurately and with reliability? To the best of our knowledge no adequate consensus trial along the lines of that carried out on our 'hermeneutic' system has been carried out on SRK. However, Kirwin (1992b) rated and assessed a wide variety of Human Error Identification techniques, including THERP, HAZOP and SRK. These were assessed by a wide variety of criteria including accuracy, validity, resource effectiveness and acceptability. SRK was found to be the worst performing of all the systems tested except 'theoretical validity' (where it was second worst) and 'auditability' (where again it was second worst). So the SRK taxonomy also fails to provide meaningful data in real world situations (Kirwin 1992).

Implications for cognitivism

It could be argued that the failure of the SRK model has implications for 'cognitivist' models in general. Cognitivism tends to propose static internal cognitive structures that stand in a direct causal relationship to external behaviours. This is its attraction, but also its weak point as more and more evidence emerges that human error is not amenable to this form of analysis. Instead, it seems far more likely that there is a dynamic, dialectical relationship between the human being and the world (Varela *et al.* 1992). If this is indeed the case, it makes problematic *any* taxonomy that uses 'cognitivist' concepts as well as raising doubts about its applicability in the specific area of safety management. Human beings perform acts which are then termed (for various sociological and psychological reasons) 'errors'. These occur in specific times and specific places. Surely the task of the safety manager is to deal with these specific situations, in terms of systems issues s/he has control over, rather than to 'investigate' 'internal' states that s/he cannot reliably identify? We would argue strongly that only by looking at human behaviour in its context (in the workplace system) will the goals of the safety manager be achieved.

Connectionism

As stated earlier, the SRK model was published in the early 1980s, at the high point of belief in the 'cognitivism' model. However, in 1986, Rumelhart *et al.* (1986) published their work on parallel distributed processors, and the debate began to move forward. The theory of connectionism (and the use of PDPs and neural nets), began as an explicit repudiation of the classic model of GOFAI, and instead attempted actually to model the neuronal structure of the human brain. PDP processing, based on the Connectionist AI standpoint,[8] is a method of arranging processors together so that they learn to perform activities. In connectionist 'neural nets', there is no 'central processor' which acts as an overall guiding force; instead, activities are decentralised. Connectionist models do not begin with 'internal' structures. They therefore proceed empirically, and build up patterns in response to external stimuli.

Once this process has begun, however, connectionist systems do not begin to build up rules about a situation. Instead, there is a 'spreading activation' in a network, consisting of forces. These forces compete with each other, and the 'stronger' force will be the one that decides upon 'behaviour'. 'Memories' of past events are co-mingled, to create generalisations which guide further behaviour.

To quote Bechtel and Abrahamson (2002):

> Connectionism can be distinguished from the traditional symbolic paradigm by the fact that it does not construe cognition as involving symbol manipulation. It offers a radically different conception of the basic processing system of the mind-brain, one inspired by our knowledge of the nervous system. The basic idea is that there is a network of elementary units or nodes, each of which has some degree of activation. These units are connected to each other so that active units excite or inhibit other units. The network is a dynamical system which, once supplied with initial inputs, spreads excitations and inhibitions among its units. ... Both connectionist and symbolic systems can be viewed as computational systems. But they advance quite different conceptions of what computation involves. In the symbolic approach, computation involves the transformation of symbols according to rules. ... the connectionist approach is quite different. It ... *does not provide either for stored symbols or rules that govern their manipulation.*
>
> (Bechtel and Abrahamson 2002: 2) (*emphasis added*)

Connectionist networks theory did not develop from 'first wave' cognitive theories. Therefore, due to their rejection of 'rule-based' models of cognition, they are incompatible with the 'GOFAI' models posited by Newell and Simon and cognitivism. If it turned out that connectionism was an accurate description of human cognition then the hypothesised levels of RBB and KBB (which rely on 'rule following') *could not* be accurate representations of cognition.[9]

At the time of writing, work has only just begun on empirical testing of connectionist models. However, it is fair to say that connectionist computer models at the moment seem to mimic human activities more impressively than GOFAI theories and cognitivist approaches. Moreover, work has now begun on modelling animal behaviour; it does seem that animal brain structures are closer to the dynamic, fluid structures undergoing constant reorganisation (Finkel 1990: 165) proposed by connectionism than the static structures posed by classic cognitivism. In any case, the point being made here is much simpler: connectionist models *may* represent an accurate model of how the human brain works, but the mind as digital computer/problem space model developed by Newell and Simon cannot model this activity.[10]

Representation

Connectionism, as we have seen, is incompatible with rule-based views of human cognition. If connectionism is accurate, then the RB and KB levels of behaviour cannot exist. This leaves the problem of symbols, representation, and, therefore, mental models.

Here it must be stressed that some of the most innovative and exciting work in psychology at present explicitly abandons representation of any sort. This is the 'dynamic modelling' approach of van Gelder and the 'artificial life' approach of Brooks (1987). For van Gelder thought is not computation but is instead a dynamic, temporal process.[11] He notes that whilst some dynamicists still see a place for representation, some have decided to abandon the concept altogether (van Gelder 1999a). The title of a text by Freeman and Skarda (1990) 'Representations: who needs them?' sums up the flavour of this approach. Freeman and Skarda argue that connectionism has not gone far enough in its rejection of 'internal states'.

We have no intention of joining this heated debate. However the key point is that to posit representation as an *a priori*, as Rasmussen does in the SRK taxonomy is surely wrong: if the concept of 'mental models' is to be used, empirical evidence must be produced as to why it is felt to be necessary. Moreover, connectionist models of representation do *not* seem to be the same kind of entity as those posited by 'old style' or 'classical' cognitivist thought (despite the fact that Rasmussen claims that the level of SBB uses parallel processing – the forerunner of connectionism [Rasmussen 1990]). So 'the cognitive structures proposed by ... connectionist networks (McClelland and Rumelhart 1986), are general, very large and universal, and *their form does not model the structure of the external world*. This view is quite different from the mental models approach ...' (Doyle and Ford 1998: 10) (*emphasis added*).

Connectionism does not imply that internal states are not present, merely that they are so dissociated and complex that studying them atomistically (and not in terms of the whole organism) is pointless. But to study the whole organism is to look at the way the organism behaves in a socially structured world. It seems that to posit 'internal' 'cognitive' states as being discrete and dissociated causal mechanisms, abstracted from the operator's social/biological system, is incorrect. And yet this is what almost all safety management 'error analysis' and HRA models tend to do.

Ecological psychology, environmental psychology, embodiment, sociation

It might seem that we are claiming that connectionism should be the basis for a new paradigm of human error, and it is true that we consider it to be fundamentally more plausible than the old fashioned 'cognitivist' view. However, connectionist models still tend to miss out two basic aspects of human nature that must be captured by any effective 'human error' model. First, and most basically, there is something that has been stressed by the 'ecological psychology' of J.J. Gibson: that Man is an animal who evolved by natural selection to exist *in a certain environment*. More specifically Gibson pointed out that it is meaningless to look at behaviour without looking at the environmental or ecological context in which it occurs. Like any animal, human beings exist in a dynamic, 'systems' relationship with the world (Gibson 1979).

Therefore, human beings always function in a specific physical environment. This is, again, not something which the 'cognitive' tradition has ignored, but its emphasis on 'internal states' has tended to downplay the situated nature of behaviour. This is something which has been stressed by the school of 'environmental psychology' (Bell *et al.* 1996). Environmental psychology stresses some fairly obvious points which can be agreed upon by almost everybody: the fact that excessive noise or heat, poor working conditions, poor ventilation, inadequate natural light and so on can influence behaviour; much 'human error' is therefore directly affected by environmental factors. But their point is more fundamental. Certain human behaviours are constrained, whilst other human behaviours are accentuated by the physical environment. If my workplace had no stairs, then I would have to use the lift, which would create new forms of behaviour and new 'error' possibilities. I can press the wrong lift button, get out at the wrong floor (and therefore become late for an important meeting) ... and so on. Much 'human error' is, therefore, shaped and conditioned by the dynamic relationship between the operator and his/her environment (c.f. Situated Cognition theory, e.g. Clancey 1993; 1997).

Realising this leads to another important fact about human beings: their biological nature. Human beings are *embodied*, and therefore human cognition is embodied too (Lakoff and Johnson 1999; Varela *et al.* 1992). The effects of sleep deprivation, stress, boredom (and so on) on the human organism make no sense unless one accepts the way that cognition is distributed throughout the body as well as throughout the environment. It may well be argued that the 'cognitivist' approach highlights and exaggerates the influence of 'higher level' cognitive processes on behaviour, while an approach that stresses biological responses to environmental pressures will produce more realistic and practical intervention strategies.

The second major aspect of human behaviour that 'cognitivism' has tended to downplay is the extent to which human behaviour is *sociated*. In a work situation there are innumerable social and physical (or to be more accurate, socialised physical) cues and 'props' to keep us 'on the right track'. For example, the door is positioned in a certain way, the computer is situated in a certain way, our workmates react to us in a certain way. These are the socialised rules and structures which govern our world. Certainly they are not rules in the sense that algorithms are rules, but that does not mean that they do not exist. If one was to break them (by, for example, attempting to drive on the right hand side of the road in Britain), one would quickly discover just how objective they were. But just because they exist objectively (in the sense we have used throughout this book, i.e. they are socially agreed upon), this does not, therefore, mean they have to be internal: 'algorithmic'. Wittgenstein in philosophy and Vygotsky in psychology have stressed the social nature of language and hence of thought, and this is particularly the case in work situations (Astington 1998). Recently, Edwin Hutchins has coined the phrase 'distributed cognition' to describe an approach to the analysis of safety and work based on the thought of these thinkers, and anthropological thought generally (Hutchins 1995).

Hutchins' idea (which is beginning to have a strong impact on human factors and safety management) is that the group, not the individual, should be seen as the basic unit of analysis for safety management. Therefore, instead of looking at 'internal' cognitive states in terms of behaviour analysis, the researcher looks at external behaviour amongst people in a group. This approach would help fill the most obvious gap in the SRK framework (and those of 'cognitivist' approaches generally), that there is no way of describing problems with communications either within or between groups.

To reduce accidents therefore, the safety manager should concentrate on changing the 'environmental' system which surrounds the operator and the organisation and societal system in which s/he has to operate. One such approach has been the PRICES error taxonomy developed by Lloyd's of London, based on a social, embodied philosophical approach (such as work by Hubert Dreyfus) with the result that '[i]t is possible to model, in a coherent way, sources of potential errors, within teams, between teams, in appropriate organisational culture, structure and managerial deficits. As little can be learnt without some degree of "trial and error", errors are a natural occurrence in this model *and have no need of explanation in terms of faulty cognitive mechanisms' (emphasis added).* (Falla 1999: ch.11, 24. See also Tomlinson 1997). This approach, therefore, is a sort of 'pragmatic social behaviourism' consistent with that posited by Activity Theory, yet another anti-'old style' cognitive approach that has been proving increasingly influential on safety management in the last few years (Barab *et al.* 1999: see links with systems theory in Blauberg *et al.* 1977). It will hopefully be seen that the 'matrix' shown in Figure 6.2 is precisely such a model of social behaviour. It is a non-causal, and non-cognitivist matrix built up from social interpretations of texts. The differences between it, the PRICES approach, and most other error taxonomies should now be obvious.

To conclude, we should mention something that lies behind everything that we have been saying without being explicitly stated: in the view of systems theory, human beings are teleological and active, not the passive receivers of information posited by cognitivism. Evidence for this, and why this is a useful paradigm for human factors is shown in Chapter ten. However, once this has been accepted, then two other conclusions follow. First, language becomes a social and purposive behaviour – a way of producing changes in the environment. And finally, given that living systems are complex, adaptive control systems engaged in circular relations with their environments then the living organism actively constructs its view of the world. It does not passively receive it. We emphasise, therefore, the *social* aspect of knowing. Living organisms make 'different observations' which can 'mutually confirm or support each other, thus increasing their joint reliability. Thus, the more coherent the piece of knowledge is with all the other available information, the more reliable it is ... there is, moreover, invariance over observers: if different observers agree about a precept or concept, then this phenomenon may be considered "real" by consensus' (Heylighen and Joslyn 2001). Hence the importance of reliability or consensus trials. They become crucial in terms of assessing the usefulness and 'accuracy' of any system that describes the world. Error taxonomies that do not meet consensus trial criteria are inadequate for the purpose for which they were created *by definition*. And if they cannot meet these criteria, they will not function adequately as they are supposed to, as error classifying taxonomies.

10 Information arousal theory (IAT) and train driver behaviour

This chapter is an attempt to answer some questions relating to boredom and stress and their relation to safety. The argument is illustrated with extracts from focus group research with train drivers. Hopefully, it will illuminate the way in which our own systems approach, which stresses the embodied nature of behaviour, differs from the 'cognitivist' approach described in the previous chapter. We have developed our own approach to operator behaviour from the theories of D. E. Berlyne (1960; 1971), who was one of the first psychologists to stress the importance of arousal as a motivating force in human behaviour, its relationship to information flow, and the relationship of information flow to behaviour.

Arousal is, in ordinary language, generally correlated with being 'excited', and is also associated with a range of biological reactions such as sweaty palms, pounding heart, perspiration and so on. While all this is going on, a part of the brain called the Reticular Activating System (RAS) also becomes more active, producing feelings of being more 'awake' and 'alert' (Apter 1992). In psycho-physiological terms, an increase in arousal tends to be associated with an increase in monoamine oxidase (MAO), a chemical which, amongst other things, regulates the amounts of serotonin and dopamine in the brain. High levels of MAO tend to produce low arousal, whilst low levels of MAO tend to produce high arousal (Zuckerman 1994). In other words, MAO functions as a thermostat, regulating bodily arousal levels, a concept to which we will return (Alverman 1999). However it must be stressed that the 'embodied' physical aspects of arousal (sweaty palms, etc.) are an essential part of the experience: it is, in other words, an intrinsically embodied state of being.

It also appears that organisms have an inbuilt urge to achieve what scientists, in a different context, call the 'Goldilocks' point: ('not too hot, not too cold'). This is illustrated by the 'Yerkes-Dodson' law that correlates arousal (termed 'anxiety' in their theory) with performance (See Figure 10.1) (Yerkes and Dobson 1908). That is, in the same way that we don't want to be too hungry or too full, or too hot or too cold, we have an inbuilt biological urge to achieve the right level of arousal (the 'optimum level') not too aroused or too underaroused; our task performance improves when this

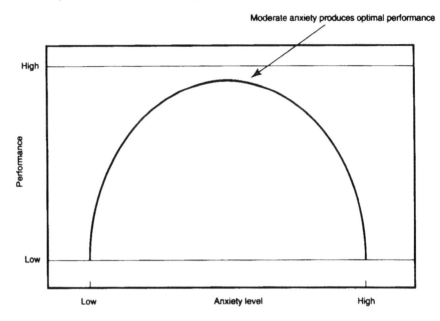

Figure 10.1 The Yerkes-Dodson law.

point is reached. Being underaroused is usually associated, in ordinary language, with being 'bored', and overaroused with being 'stressed'.

Of course, things are not quite that simple in the real world. First, the extent to which we wish to be aroused varies. In the morning, our internal arousal 'thermostat' is set at a low level, and so, for most of us, a cup of coffee and the radio in the background will suffice to help us achieve our optimal arousal level. Later on however, in the early evening, our arousal thermostat is set such that we want to be more aroused; and so we wish to go out, see a movie, go for a walk. In the same way, we wish to be less aroused when we are trying to get to sleep. It is important to understand that, for example, in the evening, we need more arousal to attain our optimal point, even though in objective terms we may actually be more aroused than we were in the morning (Zuckerman 1979).

Moreover, we sometimes want large amounts of arousal: a phenomenon termed 'the arousal jag' by Berlyne (1960). This desire is met in today's society by activities such as going on a rollercoaster or playing squash. However, for high levels of arousal, what goes up must come down: to 'complete the experience' the arousal level comes down, returning the organism to optimal again. The whole experience is perceived as being pleasurable in its totality, which means that the arousal jag has a beginning, (low arousal), a middle (high arousal) and an end (low arousal). A rollercoaster is fun for five minutes; if one were stuck on it for five days, we would feel rather differently about the matter. It should be noted that there is also some evidence that, occasionally, a moderate level of arousal above optimal can be enjoyable in itself for reasonably long periods of time (going to see a movie, for example); Berlyne termed this the arousal boost (Berlyne 1971). Nevertheless arousal will tend to return to the optimal over time (see below). The organism seeks, therefore, not homeostasis, but rheostasis: an optimum level that changes over time (Cziko 2000). There are also societal, age, gender and individual differences in optimum arousal levels (Zuckerman 1979).[1]

Berlyne's innovation was to see that changes in information available to the organism change arousal and that this can be quantified. First, Berlyne began by demonstrating a link between arousal and the 'orientation reaction'; that is having one's attention drawn to something. He discovered a link between arousal and the length of time for which something was viewed. Second, he performed experiments which demonstrated links between level of arousal (i.e. how long something was looked at) and the amount of information content of an object described in 'bits' (binary information units) as specified in Shannon's information theory (Shannon 1948). For example a straight line would describe one bit of information. A triangle (where the lengths of the individual lines were the same as the original line) describes three bits. A square describes four bits. And so on.

In one experiment, for example, Berlyne exposed adult subjects to pairs of figures, each appearing for ten seconds, and measured how long they were looked at. It was found that more complex figures (i.e. figures that contained more information) were looked at for longer, and therefore were more 'arousing' than simple figures. In a similar experiment, Berlyne showed that one other criterion was important (in rats as well as humans!), namely 'surprisingness'. 'Surprisingness' is basically the way in which one piece of information differs from the last piece; in a sense how 'unexpected' it is, given what has gone before (e.g. 'God save our gracious *hamster*'). He also showed that a constant flow of information will, other things being equal, tend to be more arousing in its totality than an infrequent flow. If we went to see a movie which played at one frame per twenty minutes, it would be a good deal less arousing than one playing at the normal speed (Berlyne 1960). The corollary of this is that given an infrequent flow of information, a sudden increase in information will be 'surprising': for example a loud noise in the context of silence.

Therefore, information flow which is complex and continuously varied ('surprising') will maintain arousal, and, by contrast, information flow which is simple and monotonous ('redundant') will lower arousal and cause a state of boredom. At this point it is worth noting classic studies of the type carried out by Vernon (1962) in which human subjects were kept in a 'completely homogenous and unvarying environment', that is a virtually information free environment (170). Subjects suffered from extreme boredom and restlessness, deteriorating IQ and even hallucinations. Few were able to stand the experiment for more than two days, despite being extremely well paid. Boredom (information underload) as well as stress (information overload) thus has serious performance consequences. In that sense, often-heard management statements of the type, 'It's easy. All s/he has to do is sit there and press the button when the light comes on. How can you possibly get that wrong?' seriously misunderstand the nature of human performance under conditions of low arousal.

A final point should be made. In later experiments Berlyne discovered that when asked what people *liked* or *disliked*, in the same way that they disliked over-simple, boring shapes, they also disliked over-complex, 'over-surprising' shapes. For example in an experiment, subjects were asked to rate complex patterns in terms of how much they liked them. Over-complex patterns were less liked, as were over-simple ones (Berlyne 1974b), and Berlyne demonstrated that this was because of the link with arousal. In general (with the provisos mentioned above) too much arousal is unpleasant, as is too little.

People: controllers of arousal

Although in information terms we have used the language of discrete 'bits' to quantify (digital) information, it is highly likely that in a 'real world' situation humans perceive information in an analogue fashion. That is, people see things as being more or less complex or surprising, rather than in a digital complex/not complex or surprising/not surprising manner, and that therefore the language of 'fuzzy logic' is more appropriate than the 'digital' language used by Berlyne ('fits' not 'bits': Kosko 1994). It is difficult to describe the *Mona Lisa* or Beethoven's Fifth Symphony in terms of 'bits' without the exercise becoming increasingly arbitrary and eventually rather absurd (see Moles 1968). However, this does not alter the basic thrust of the theory.

The chase for arousal is an example of the brain acting after the fashion of a thermostat, as discussed by Ashby (1956). Operators appear to act not as passive beings forced to be aroused or not aroused by the external world, but as active regulators seeking out the ever-changing preferred arousal point: rheostasis. It is as if they were saying, 'Do I like this change in information flow? And if not, how shall I change things so that I do?' Information is gathered (feedback) which alters their 'system' and this has a hedonistic/evaluative (preference) component such that the person will then seek out more information or less, depending on how close it is to the rheostatic point. Operators are part of the system; permanently and actively part of a feedback ('cybernetic') loop (Wiener 1949). However, not only is this arousal point constantly changing (not just in terms of rheostasis, but also in terms of 'the arousal jag' and the other variables discussed above), but the nature of the information flow also tends to mitigate against keeping to 'optimum'. The catch 22 is that when one has reached the ideal kind of mixture of complexity and surprisingness it gradually (or rapidly) changes, because the *same kind* of information becomes boring for the simple reason that the *subjective information content* declines with repetition. The incoming information must therefore be *dynamic* to maintain arousal.

For example, say one is in an art gallery to look at the paintings. We need arousal to function, and so we seek out a picture that strikes us as 'interesting'. However, after a while looking at it, the same information at the same rate (i.e. constant) leads to lower arousal. So we move off and look at another painting and then the process repeats. Only constantly varied information can keep us at the optimum level, but because only we can say how arousing the information is (because only we can compare it with previously experienced information), only we can decide what information, and how much, will keep us at the optimum (Lynn 1966).

Therefore we need to have *control* of the information in a situation to prevent boredom. We do not passively receive information, we actively go out and seek it: and not just any information but the kind we need to increase or lower our arousal. There are obvious links here with the theories of J.J. Gibson, who also posited humans as active, information seeking organisms, and like Gibson, we would emphasise that the incoming information is not 'cognitively' processed, but instead 'resonates' with the whole organism (see Gibson 1979). This fits in with the current views of neurophysiologists who find no evidence of bit by bit 'information processing' in perception (Freeman 1997; see also Clancey 1997).

There are also links with the theories of John Adams (1995) and Gerald Wilde (1994), who have both posited human beings as possessing a 'risk thermostat', which must be adjusted to maintain adequate levels of arousal.[2] Human beings seeking risk and human beings seeking information may seem to be different hypotheses but as we shall see there are links between the two theories (see also Zuckerman 1979 for links between arousal levels and risk taking).

Thermostats

The idea of human beings functioning as thermostats is one of the best understood in systems science. To understand what is meant by it, take the example of a real thermostat, in a real room. This is part of an *open system* (because the house is open to the external weather/ecosystem). Technically all systems are open systems, but in practice self-enclosed technical systems can sometimes be considered as closed systems. All systems involving human beings, however, should be considered open systems.

In this open system, therefore, the thermostat is set at a certain level. This is the temperature which is desired: the optimal point. The thermostat has at its control a heating mechanism with which it will attempt to attain this optimal point. Information is received by the thermostat (from an external thermometer) which will enable the thermostat to raise or lower the temperature to the desired level. Information comes to the thermostat (in the form of readings from the thermometer) which is then acted upon. This action then alters the system (i.e. the temperature), and the results are 'read' back in the form (again) of information, which again then leads to actions by the thermostat. This, therefore, is a teleological attempt by the thermostat to control the system and achieve homeostasis.

Now, the interesting point here is that once the system is up and running, strange and non-Newtonian things appear to happen to causality. For example, what 'causes', the temperature to change in the room? Obviously the thermostat. But what caused the thermostat to act and change the temperature? It was information through the medium of the thermometer. But what produced this information? The temperature of the room. And what caused the temperature of the room to change? Amongst other things (but primarily) the thermostat.

In other words, the cause (thermostat) produced an effect (changes in the room temperature) which then acted as a cause (on the thermostat) producing an effect ... and so on and so on. This is an example of 'circular causality', an inevitable feature of complex systems using feedback. The cause produces an effect but the effect is also a cause, producing an effect and so on. It is obvious why, given this non-Newtonian view of causality (Newtonian causality being seen as simple 'one-way' 'cause and effect' relationships, or, in psychology, stimulus-response), and the complexity of the causal loops which are involved, systems theorists prefer not to use the cause and effect metaphor (Lakoff and Johnson 1999), instead talking about systems which are in or out of control (Flach 1999). For a stronger view see McCullough, who claimed that 'causality is a superstition' (McCullough 1965: 148).

Another conclusion which follows from this 'systems' view is that the operator and the system cannot be considered meaningfully as separate units. It is futile to ask why the thermostat acts as it does without discussing external room temperature, and pointless to ask why the room is a certain temperature without discussing the role of the thermometer. Therefore the subject (the thermostat) and the object (the room temperature) are not distinct; instead there is a dynamic system with a teleological component (in this case, the thermostat) regulating the system using constant feedback from within the system (Varela *et al.* 1992 call this *structural coupling*). The dissolution of the 'subject–object' distinction is also present in Gibsonian Ecological Psychology (Gibson 1979).

Another conclusion follows on from this last point. If an operator is within a system, and is therefore part of the system, and is behaving purposefully, then s/he is seeing everything from a point of view. In other words, s/he never sees anything 'disinterestedly' before it is 'processed' internally; instead the world is perceived in so far as it can help in the control task s/he is engaged in at the time. In fact this is true *a priori* given the abolition of the 'subject-object' dichotomy: the purpose for which an object is perceived will affect how it is perceived. This perspectivism fits in not only with systems theory but also with contemporary epistemology (Davidson 2001).

A final point must be understood: at no point do we need to hypothesise either a mental model of the external environment or rules which evaluate relevance of information or action. Instead, the embodied operator engages dialectically and dynamically with the 'external' environment of which s/he is a part. This is, therefore, not a 'cognitivist' theory of behaviour.

Talking about boredom

People sometimes adopt surprising ways of regulating information flow, and thus arousal. This was revealed in the natural conversations obtained during focus group studies carried out with train drivers during 2000 and 2001. The focus groups were held in order to shed light on why a particular type of train (or 'set') was involved in passing red signals (when this happens, a SPAD is said to have occurred: 'Signal Passed At Danger') more often than another type of 'set' (NB: the differences in mileage travelled by the different sets was taken into account). Immediately, initial thoughts turned to technological or engineering reasons, particularly braking systems, as an explanation for this state of affairs. However, the focus groups suggested that drivers perceived no inadequacy with the brakes. What they did suggest, however, was that this type of set was mainly driven on routes that they found *monotonous* or *boring*. Therefore, on these routes at least, boredom appeared to be associated with a higher SPAD rate. Information Arousal Theory (IAT) offers clues as to why this might be.[3] (Report commercial in confidence.)

During the focus groups, drivers appeared to link boredom with fatigue and lowered performance (for links between boredom and fatigue see Mackworth 1969). For example one driver commented:

DRIVER A: Repetition sometimes. You maybe get it more. You get sent on the same run ...

DRIVER B: [... names some routes] the repetition there was horrendous.

DRIVER A: And what you find is you only react, you know, you could go round the circle, come back in and not remember doing it, because nothing's happened.

DRIVER B: You're on automatic pilot.

DRIVER A: It's the monotony, just going round and round and round.

One driver also pointed out that the 'problem' sets tended to be used as a 'workhorse' for the more tedious sort of jobs, again suggesting that the higher SPAD rate on this set was not due to technical issues, but instead could be better explained in terms of information and arousal.

This link between boredom and error has been shown in empirical studies (McBain 1970[4]; see also Edkins and Pollock 1997 for specific data regarding train drivers). However, instead of passively accepting the monotony, drivers described a number of ways in which they attempted to increase or vary information flow in order to increase their arousal. So, for example, drivers ended up daydreaming, playing games (such as, literally, counting sheep), directing their attention towards things not relevant to the driving task, 'chasing signals' or even singing out loud in the cab!

DRIVER C: You feel isolated, very very isolated.
DRIVER D: Especially see if you're a person that likes to talk ...
DRIVER C: ... You end up singing! Sing a different song! It's monotonous.

The problem thus appeared to have nothing to do with the 'problem' sets themselves, but everything to do with the way they were used; the drivers tried to compensate for low information/low arousal by adopting personal strategies. Some of these strategies including unwise acts or even rule-breaking, but arose as a consequence of the low-information regime imposed on them. This type of finding has been found in other studies of boredom; operators attempt to manipulate their information flow to maintain arousal (O'Hanlon 1981; Fisher 1993).

Talking about stress

However, it also emerged from the discourse that the opposite situation could arise when drivers begin their careers. They often feel they are given insufficient training, are not as well equipped to deal with the demands of driving as they would like, experience information overload, too much arousal, and therefore become *stressed*.

DRIVER E: I think that the kind of training we got, that was totally inadequate to be honest with you.
DRIVER F: We didnae really get long enough.
DRIVER E: Unofficially it's been well known throughout the system that [...] shocks, you know, you get a shock whenever you're going like, and you end up with this experience [...] nowadays kids just get chucked in at the deep end basically.
DRIVER F: It's just pressure at the end of the day, it's the pressure once they take that set away [...] it could be a hundred and million daft things, you know when you're panicking, it could be a simple thing, but that's the thing that you miss, you know, because you're worrying about everything else, and you're panicking, you know.

This, therefore, is stress resulting from information overload.

The arousal thermostat

It should be emphasised that much work has been done on the relationship between stress, boredom and arousal. Moreover, much work has also been done on the relationship between information and arousal. However, the key concept that has, perhaps, been under-emphasised in previous studies, is that the operator functions in a systems loop with the environment in order to achieve the desired level of arousal. This is where the theory propounded here fits in with the concept of the risk thermostat as discussed by John Adams (Adams 1995) and Gerald Wilde (Wilde 1994). According to the focus groups, when drivers are bored (under-aroused) they will manipulate the environment in order to raise their arousal. This can be done by increasing the information level available to them, but another key way of increasing arousal is to increase one's risky behaviour (producing the 'flight or fight' mechanism). This is not an either/or situation, since everything can be conceptualised as information. However it *is* clear that drivers who sing, daydream, 'chase signals' or other personal strategies are attempting to raise their arousal level. On the other hand some drivers see the need to reduce arousal levels.

DRIVER G: I think they should have some chill-out rooms in every depot, a little bit of ... er ... [5]
DRIVER H: ...unwinding.
DRIVER G: Background music, dim lights, and all this like.

What they are asking is for the opportunity to be able to lower their arousal when they are overaroused (stressed). Once again, drivers seek control over their arousal level.

Control

The concept of 'control', a key concept in systems theory, may be of help when attempting to gauge when arousal is pleasant or unpleasant. When people feel themselves to be in control of a situation they can tolerate 'unpleasant' situations because they are controlling the situation themselves (Weiss 1972; Glass and Singer 1972). Stressful situations, on the other hand, tend to follow when the driver feels s/he has no control of the situation: in other words, s/he cannot lower (or raise) his/her arousal level him/herself. This is particularly important as continued lack of control might lead to a situation of helplessness in which the operator 'gives up' and stops even attempting to concentrate. Perhaps even after control has been returned to the operator he might remain in this state of helplessness ('learned helplessness'). Experimental evidence has demonstrated that a state of learned helplessness is correlated with lowered performance. The implications of this for a train driver are obvious (Hiroto and Seligman 1975).

Most up-to-date theories of human action stress the teleological, control-driven aspect of human behaviour. This is important in terms of man–machine relationships. If drivers are in control of the information available to them they can regulate their own arousal levels, but if they cannot do this stress or boredom result, both of which, as the drivers acknowledge, can lead to unhelpful or even maladaptive coping strategies, mistakes and eventually accidents.

Therefore emphasis should be given to strengthening and increasing the amount of *control* drivers have over their arousal/information levels. However, they require the means to do this in ways that increase, rather then decrease, safety. Their suggestions included the following.

DRIVER I: You're meant to drive and concentrate, that's your job, but there must be something could be done to stimulate you …
DRIVER J: … make it easier to keep concentrating.
DRIVER I: You get repetition because of the nature of the job, that's the same as a pilot …
DRIVER J: … he's got a co-pilot though hasn't he …
DRIVER I: I do think we should have cabin crews coming in and out all the time.
DRIVER J: I do think maybe they should have some kind of radio, I think to relieve the boredom I think, 'cos it's really mentally tough – going on your own for ten hours, and I think –
DRIVER I: A radio, I mean it's not just …
DRIVER J: But I really think they should look into something about that, you know, the boredom factor, do something to take it away …
DRIVER I: The bus drivers have been doing it for years, driving about with the radio on.

Whether their specific suggestions have merit is beyond the scope of this chapter to discuss. However, the desirability of helping them control their arousal levels, especially where the natural affordances for so doing are very limited, echoes a point made by Reason (1974: 148–9):

When background stimulation is low and signals few and far between, the level of arousal – and with it vigilance – begins to decline because the quantity and variety of the sensory inputs reaching the reticular formation are insufficient for it to carry out its job of maintaining cortical efficiency. […] Interestingly enough, many of these antidotes to vigilance decrement were unwittingly present on the footplate of the old steam locomotive. In between shovelling coal, the fireman would sit opposite the driver and help him with the identification of signals, train speeds and other problems connected with driving the engine. […] it is obvious that this system carries with it at least two advantages. Firstly, it kept the driver alert through the additional auditory stimulation and the presence of someone else in the cab. Secondly it provided the driver with confirmation of any unexpected signal. [...] unfortunately the advent of the diesel engine broke up this happy partnership [...] many trains began running with only one man in the cab.

This is something backed up in the focus groups.

Further implications

Going back to the 'thermostat' idea mentioned before, we should remember that that was an example of an open system, and that all systems should be considered as open unless it can be shown to be otherwise. Now, in that example, the temperature of the house was affected by the external weather. This can be represented as D (Disturbance). This has to be counteracted by the thermostat (R for Regulator). Now, say that D can function in two ways: A and B. Function A is to lower the temperature and function B is to raise the temperature. Now, if the regulator can only use function A, it can only control the outside temperature if it goes up. It needs both functions even to attempt to control the temperature adequately (considering outside disturbances might drive the temperature down or up). So if we define disturbance A as 'external temperature going up' and disturbance B as 'external temperature going down', the regulator must be able to match these disturbances by responses of its own. So it must be able to provide as many responses as there are disturbances in order to keep the temperature under control (so in this example, R can match disturbance A with function A and disturbance B with function B). This can be illustrated in an equation termed Ashby's law of requisite variety:[6]

$$V(E) \geq V(D)-V(R)-K$$

Where E is the Essential Variable (in this case the rheostatic point), and K is the effect of 'buffering': the unavoidable lessening of disturbances – for example in the thermostat example the walls of the room would be acting as a buffer, lessening the effect of the weather (Ashby 1956; Heylighen and Joslyn 2001). To return to our drivers, they need ways of reducing boredom *and* of reducing stress. If these things are not planned and designed for, they will produce their own solutions!

This chapter has concentrated on the arousal control mechanism, which, we argue, is a 'feedback loop', and, therefore, a situation where the 'cause and effect' metaphor is not relevant. However, it is important to understand that in this view, all human behaviour is, at least in principle, perceivable as a series of control mechanisms.[7] These mechanisms are usually arranged in a hierarchy with some predominant at some times, and some at others. When discussing this aspect of behaviour, systems theorists have emphasised morphogenesis rather than homeostasis: the subject's activity towards new goals based on positive feedback, rather than back towards the old goal based on negative feedback (Geyer and van der Zouwen 1991). However, the basic concept remains: organisms self-regulating via positive or negative feedback.

It is also clear that the better the subject can predict the 'consequences' of an action, the better the choice will be as to what action to choose. In other words, in informational terms, the subject should attempt to gauge the probabilities (or fuzzy certainties) of actions. This can be described as H(R/D), where H is the degree of the subject's knowledge of social heuristics which facilitate or constitute action. Therefore the equation can be modified:

$$H(E) \geq H(D) + H(R/D) - H(R) - K$$

which merely states the fact that decreasing uncertainty (or, in ordinary language, increasing knowledge) will also facilitate moving towards the desired state.

This is very important in that it seems that many accidents are in essence unpredictable (Wagenaar and Groeneweg 1987) and are only blamed on 'human error' because of attributional biases and the convenience of the term 'human error' as an explanation. Despite all our best efforts, it still makes it look like someone's fault. But to what extent are these types of accidents really 'human error', or really 'systems problems'? We have argued that this is a meaningless question for two reasons. First, there is no distinction between the subject/object–human/system. They are part of the same intertwined whole, with the human as the 'teleological' partner endlessly manipulating the 'external' environment. Second, attempts to 'attribute' blame to one or the other will always be constrained by attributional biases.

If this is the case then some of the 'solutions' often proposed to solve safety problems become themselves problematic. For example, Wilde (1994), claims that, from a systems perspective, perhaps we should reward drivers if they have less SPADs. After all, drivers are already punished for having 'too many' SPADs: why should they not be rewarded for having fewer? Wilde also argues that if people seek arousal, the very fact of having broken a rule may cause the arousal: in other words, 'strict rules' in the wrong context may lead to an increase in rule violations, due to 'the buzz factor'! He cites a Swedish study in support of this worrying suggestion, in which car drivers who had obtained a relatively severe punishment for drunk driving were compared with a second group, who had received a milder punishment. The heavily punished were *more* likely to reoffend (Wilde 1994: ch.11, 3).

However, many of Wilde's examples involved car drivers, who, almost by definition, have a huge amount of control over their environment. They can listen to radios, drive where they want, and so on. Train drivers are in a very different situation. For example, when a train driver feels sleepy, realistically he may have no choice but to continue driving. Moreover, as one might expect in a systems context, the 'blame' context that may arise when staff are rewarded or punished has its own feedback problems: specifically low reporting rates. Working out 'who was responsible' is of course also problematic in attributional terms.

So while we are not in theory against this idea, we would emphasise that it must take place in the context created via Ashby's equation. Most importantly, drivers must feel they have control over *their system*. Furthermore, even *subjective* feelings of control can help to improve performance, showing the importance of creating a culture where drivers and operators generally feel not only that they have control over their environment, but that they can get adequate and meaningful feedback on performance (Glass and Singer 1972; Weiss 1972); that is, that their complaints, thoughts or suggestions are valued, acknowledged, taken seriously, and acted on where appropriate (job satisfaction, or lack of it, functions as a SPAD predictor, see Edkins and Pollock 1997).

Control and technology

Much of the development of automation has had the effect of removing control from the operator, in favour of computer systems and 'fail-safe' mechanisms. By contrast, we argue that in order to regulate arousal (and, therefore, increase vigilance) the driver should have *more* control over his/her environment, specifically as regards information. This suggestion is borne out by current research on air traffic controllers (Metzger and Parasuraman 2001). It is interesting in this respect that a Swedish study of railway cab conditions stated that 'not enough information (is currently) provided to the driver' and that when information is made available it should be 'dynamic information' (Lecklund *et al.* 2001).

To be done safely a job should be *interesting*. This is a greatly neglected topic in ergonomics (Fisher 1993; Smith 1981). However, automation, by reducing the complexity of the task, may contribute to making it more boring, surely another 'irony of automation' (Bainbridge 1987). As Rudisill writes in her survey of airline pilots' attitudes to automation: 'There was a general concern that automation may increase boredom, thereby indirectly decreasing safety' (Rudisill 1995: 3).

Causality, consensus, cognitivism, cybernetics

Finally, there are what one might term 'philosophical' implications, which are, in many respects, the philosophical conclusions of this whole book. First, there is the support within systems theory for the 'perspectivist' approach. We always see things from a point of view. If this is the case, then, how do we build 'a picture of reality'? We build it via consensus (triangulation) such as we have discussed in Chapter eight (moreover this emphasises the extent to which Heidegger's hermeneutic circle is a systems 'feedback loop'). We engage with the text and gain information and on the basis of this information then produce new readings and so on. We always see things from a point of view.

Second, given that cause and effect are irrelevant in this particular 'systems' discussion because of the 'thermostat' analogy, we see why attributions like those discussed in Chapter seven are produced. People are being asked the wrong questions: questions about causality where causality is not the best language to use. From this viewpoint, questions like 'who was to blame' or 'what caused the accident', or statements such as '94 per cent (or 88 per cent or 92 per cent or 70 per cent) of accidents are caused by human error' are literally *meaningless*. The language of Newtonian causality is simply inappropriate in this context (Flach 1999).

It should also be obvious that this is a socially constructed approach, that takes account of social relations between levels of an organisation as shown by the language used by the reporters, and the differences in their attributions and accounts in different settings (e.g. in a confidential focus group, as opposed to a management meeting).

Finally, we have demonstrated an approach to human error that makes no use of cognitivist concepts such as 'algorithmic rule following' or 'mental models'. Instead, we have the socialised, embodied operator attempting to manipulate the environment in which s/he is embedded, and telling us how they seek to do this. We suggest that this is a useful paradigm for safety management, and more modestly that some of this thinking has implications for cognitivist paradigms in psychology in general. As Hollnagel (1983) has argued, 'human error' is a meaningless concept. It is only through a teleological systems approach, we believe, that real progress can be made to increasing safety.

11 Conclusions

The book is about science, and about language. It proposes ways of dealing with natural incident ('in your own words') reports in a way that extends the normal qualitative methods currently in vogue, turning reports into the kinds of data necessary for those working in high-risk industries whilst preserving the philosophical status of language as performative and symbolic. We have in the process of our discussions looked in some detail at the nature of language and the ways in which the words we use to describe things and situations both reflect, but at the same time, give rise to, our impressions of the world and the ways we understand and construct it. Since this process is highly individual, with no two people constructing things in precisely the same way, dealing with language from a naïve/realistic stance whereby words, rather than being performative, simply mean what they denote, creates major difficulties. The problem of 'what to do with' reports of incidents and near misses, and of deciding 'what people really mean' illustrates this problem nicely.

In the past, the general trend has been towards asking an 'expert' to interpret what the import of such reports 'actually is'. An engineer or other suitably qualified person reads the reports, decides what they mean, and perhaps ticks some boxes in an engineering taxonomy; or else, maybe after some head scratching, puts a tick in a box called 'human factors', or equally unhelpfully, 'other'. Taxonomic problems aside, on the surface this looks fine but makes two assumptions about experts that are simply not true. First, that they will always agree; second that their judgements are for some reason free of the subjective biases and motives that afflict lesser mortals. We have reviewed the evidence, and conclude that neither of these assumptions is true.

If this conclusion is a reasonable one, and we believe that it is, then the interpretation of text about near misses, minor incidents, and safety concerns by experts in engineering, in a particular sense, has something in common with hermeneutics, which originated in the study of religious texts by priests. Note however that the emphasis here is on the process of interpretation, of extracting meaning from text. We are not saying that safety text analysis is a metaphysical enterprise based on Divine inspiration;

merely that both priests and engineers are involved in the process of extracting the important messages from what are fundamentally historical documents.

It has frequently been assumed that experts have context-free and unmotivated interpretative abilities that give them access to the true meaning of the text. However, as noted previously, experts often disagree, frequently along fracture lines that are entirely predictable when something goes wrong, litigation is involved, or you know which 'side' they are on! In such cases, the view that prevails is usually the one decided by a consensus of some type; either of other experts via an inquiry, a lay jury, or some other means – a process not very different in principle from the system proposed here for dealing with the text of incident reports. The view of the objective, dispassionate expert pursuing some knowable truth is replaced by the view of the expert as the person possessing certain knowledge, who uses that knowledge to defend positions, attack others, decide what is reasonable and possible within economic and political constraints, decide which causes to fix, and pay the mortgage. Furthermore, the only way to escape from such a minefield of potential or actual idiosyncracy is via consensus.

We have argued that a pragmatic view of science gives up the pretence of the objective expert, and takes on board views of science (already existing) that see the investigator or researcher as a dynamic component of knowledge itself, in interaction with the stuff of the universe, selecting, interpreting and at the end of the day having a material impact on the form that knowledge takes. That is the way things are, and the pretence of a certain universe, providing certain and absolute truths discoverable by zombie-like experts with no motives or interests of their own, is blatantly *unscientific*.

Numbers from words

What we have attempted to do in this book is provide a method for dealing with the natural discursive accounts that people provide when things go wrong, in such a way that there is, firstly, agreement about their meaning. Agreement about meaning then permits their classification in terms of taxonomies of human action and human error that situate such actions and errors in the specific contexts within which they occur within specific industries. From that starting point, the interpreted (coded) texts can then be treated in exactly the same way as the outputs from any engineered system, in terms of trends, graphs, and particularly control charts. The first specific conclusion, therefore, is that there exist methods (we prefer the term 'methods' to the current vogue of calling such things 'tools')[1] which can turn natural reports into the type of output that can be used and analysed in the same ways as 'hard' data. There is no need, from this view, for a separate and subordinate category of 'subjective' evidence whose epistemological status is more doubtful than 'hard' data, which cannot be turned reliably into numerical form, whose only function is to shed light on 'better' data when the opportunity arises to do so, and which may be ignored when such a match fails to materialise. We have presented evidence and case histories from our own experience showing how verbal and textual reports can complement other information on equal terms, and sometimes suggest lines of inquiry which are not suggested by a straightforward engineering approach. Sometimes the solutions to such issues are more effective and cheaper than the latest technological fix.

Perhaps the first major point then, is that dealing with natural safety reports requires some basic familiarity with the nature of language, and the skills necessary for dealing with reports written in the person's own words. This is not rocket science, but it *is* a body of knowledge that requires to be known, if the maximum benefit is to gained from the analysis of natural reports and if mistakes are to be avoided. Obviously, and self-evidently, engineers have to know about the methods and philosophies implicit in dealing with engineering issues and, equally self-evidently, people dealing with natural language need to know something about the methods and philosophies required for that task.

The method we suggest is simple, but may take a little getting used to. To begin with, it requires reporting systems which encourage reporting, despite the view in some quarters (in our own experience) that safety reports are bad news, and that their absence is good news. We have discussed the merits of confidential reporting systems, particularly those run by independent third parties, and argue that a combination of these two factors (anonymity, independence) can stimulate the numbers and types of reports required by a system of the type suggested. Where safety is concerned, with the possibility of loss of life, injury, and resultant litigation, we would argue that it is time for independent firms and companies to leave behind the secrecy and seeming paranoia with which some of them view their safety data. Where business matters are concerned, competitiveness requires that companies guard certain relevant secrets, keeping themselves to themselves. Safety matters, however, do not ethically belong in this category, so that systems such as CIRAS become, hopefully, a vehicle for change and improvement in safety from which an industry as a whole should be able to learn, as well as individual companies. Arguably, and in an ideal world, *all* safety data should be available on a national industry-wide basis. However, the world is not ideal and in some industrial sectors there exist tensions between company-run, industry-run, and independently run safety databases, with some companies preferring that their safety data should remain very much 'in house' for reasons which to us sometimes seem to have more to do with economics than safety improvement.

Whatever the truth of the matter, the system we propose requires reporting on a reasonably large scale. This again may seem counter-intuitive, for two reasons. First, it might be argued that in a well-run company where safety is excellent, there should be nothing to report. Second, in existing systems, we sometimes hear expressed a concern about the number of reports received which appear to be either 'trivial', or 'gripes and moans' of minor safety significance. It is worth looking at these concerns in a little more detail.

The apparent common-sense belief that a well-run company should produce no safety reports is misconceived. No organisation is perfect, and the barriers and defences in place can never be so absolute that accidents or unwanted incidents are impossible. Knowledge is never complete, and certain types of prediction are in principle not possible as we saw from Chapter two. To prevent everything is impossible and invites financial suicide if one should even attempt the task. Consequently, safety depends on a continuous flow of information from staff in order to identify where weaknesses occur in systems, and where new weaknesses emerge as systems evolve and change. For example, where fundamental change is proposed (e.g. reducing staffing levels) a healthy reporting system can be a crucial safeguard against the type of 'precipice culture' problem described earlier, and can identify weaknesses consequential on the change, whatever its nature. However, the extent to which a reporting system is actively used has far more to do with management's attempts to encourage (or discourage) open reporting, their honest desire to know what's really going on, their response to the reports received, and their desire to share the lessons learned, than with some intangible measure of 'how safe' the company is. Whilst a regular flow of reports is no guarantee that a company is 'safe', the evidence generally suggests that companies producing no such reports have a poor safety culture and poor safety awareness. So whilst a healthy stream of safety reports (in itself a desirable feature of a safety conscious company) does not guarantee 'safety' (however measured), an absence of such reports is a very bad sign indeed where it indicates a management style that simply 'doesn't want to know' and discourages reports from people it sees as 'trouble makers'. Our own experience suggests that this produces a 'keep-your-head-down-and-say-nothing' attitude amongst frontline workers, a form of adaptive learned helplessness, which is inimical to any form of safety improvement. A healthy reporting system, therefore, is a way of involving staff in safety issues, maintaining vigilance and safety awareness, and can be a component of that illusive entity known as safety culture.

However, there is another reason why a regular stream of safety reports is necessary, which is much more tightly defined. We have seen in Chapter three how the assumption, that the causes of minor events and near misses at the base of 'triangle' models are also represented amongst the causes of accidents and fatalities at the top, is justified to a useful extent. Therefore, tackling the causes of minor events can be expected to impact on the frequency of major events. However, the difficulty lies in knowing which of the much more numerous minor event causes are the ones to tackle, since tackling them all is simply not cost effective. The answer to this problem constitutes one of the main purposes of this book, namely how to separate signal from noise in the reports received. In other words, how to separate out the regular stream of reports into those that are trivial or just 'gripes and moans' from those that require action. It goes without saying that in order to separate out signals from noise you have to have some signals and you have to have some noise.

It will be recalled from Chapter six that control charts can be computed from incident reports, in a way which permits the identification of periods of time during which reports of a particular type can be said to be 'out of control'. From a statistical point of view, the modelling of this type of data becomes more useful and more robust the greater the number of data points to hand. Data points are, in this instance, reports. To put this in the simplest terms, if one is attempting to identify times when reports of a particular type are being received at a rate which is abnormal, the accuracy of the model depends to a large degree on how good a measure is available of what is the normal rate. Furthermore, the 'normal rate' is continuously re-calculated as more data are gathered, so the estimates become more and more accurate. To identify abnormal variation, you have to be able to *measure normal variation*, and keep revising that measure over time. A regular stream of reports of all types, including those that may initially seem to be just 'gripes and moans', is required to form the 'background' against which abnormal reporting rates can be identified.

Finally, on this topic, it only remains to emphasise one last time that reports are required not merely of serious events, but of minor events and near misses also. In a sense major incidents report themselves. The problem to be addressed is the usefulness of minor event reports, which appear to pile up in filing cabinets, often to no great apparent purpose. However, we refer again at this point to Chapter three, on the 'triangle' models of accident causation. To be useful, the 'base' of the triangle has to be filled up, and the base level is largely constituted of the type of minor event reports we are talking about. Perhaps even more importantly, the base of the triangle(s) is where the clues to accident and incident *prevention* lie. By definition, at the higher levels the things that should have been prevented have already gone wrong. Prevention thus requires, and to a great extent survives upon, a healthy and regular stream of minor-event reports.

Reliability

The point was made in Chapter eight that the process of coding reports into any sort of taxonomy has to be carried out reliably, and we were at some pains to clarify our own meaning for the word 'reliability', and to differentiate it from physical or engineering definitions of reliability as applied to materials or machines. Basically, the fundamental requirement for any coding system of whatever type, is that people (coders) will assign the same codes to the same events in the same way, on different occasions. For example, suppose I give an engineer a file containing the evidence about an event, and ask him to code it into a taxonomy. It would be my reasonable expectation that if I waited a few weeks (to give him time to forget the details and the decisions he had previously made) and then asked him to repeat the exercise with the same event file, he would code it in the same way. If he did, he would show 'intra-rater' reliability; that is, he would show *consistency with himself* in the task of coding. However, it is impractical and probably impossible for one person to code all the reports for a firm or company, let alone an entire industry; more than one person would normally be required to handle this job if the reports were coming in at the rate desired. To show *consistency with others*, that is 'inter-rater reliability', I would need to give the same event file to the different people involved in the coding task, ask them to code it into a common taxonomy, and then look to *see how much they agreed with each other* in terms of the codes they assigned. In simple terms, where events are coded into a taxonomy by different people, there are three requirements if data are to be comparable and capable of being added together so that broader, possibly industry-wide, lessons can be learned. First there has to be an agreed taxonomy that is unambiguous and *capable of being used reliably*, about which more will be said in the next section. Second, each coder has to code events consistently over time, and third, that consistency has to extend to any others involved in the coding of events. Only then can the data from events be taken as having discriminative and predictive utility with regard to the events coded. By contrast, where there is little consensus in either of the above senses, any database (other things being equal) simply describes the different general biases, beliefs and coding preferences of those doing the coding, diluted by the extent of mood swings of individuals on different days. In the extreme case (admittedly a *reductio ad absurdum*) a discriminant analysis of a database where variance was primarily due to between-subjects (inter-rater) variability, would simply shed light on *who* coded which events or, in the absence of within-subjects (intra-rater) consistency, nothing at all! The point is, however, that in the absence of reliability trials and data, it is completely impossible to tell to what extent a database of assigned taxonomic codes describes a consensus and agreed ('objective') version of events that took place, and to what extent it merely represents the idiosyncratic views and personal prejudices of the person or persons who do the coding.

The above discussion leads naturally to consideration of statistical methods for measuring reliability, and a survey of the literature reveals considerable disagreement and inconsistency in how best to go about this. Accordingly, in Chapter eight we recommended a Bayesian statistic (BK) (Lee and Del Fabbro 2002) for determining the extent to which coders can agree on the categorisation of events. We argued that this seems preferable to using the traditional 'frequentist' Kappa Coefficient (Cohen 1960) which can be shown to vary independently of agreement.

Computing reliability

Whilst this is not a book about statistical approaches, it is worth noting that there seems to be much to the Bayesian approach in general which lends itself to the type of science we are proposing. Suppose we are interested in whether a factor (say fatigue) is associated with another (say accidents). We might compute a correlation coefficient between measures of fatigue and frequency of accidents, and then test the significance of the 'r' value obtained. In traditional 'frequentist' inference, these tests of significance are performed by supposing first that there is *no relationship* between fatigue and accidents (the null hypothesis) and then computing the probability of obtaining the data collected if this was the case (this is the *P-value*). In other words, frequentist statistics examine P (D/H): the probability of the data given a 'true' null hypothesis that there is no relationship between the factors under examination. There are two crucial aspects to this which have been criticised, and which are addressed with a Bayesian approach.

First, it has been pointed out that the null hypothesis is almost always known to be false from the outset (Edwards *et al.* 1963: 236). Do we really imagine there is *no* relationship (i.e. a correlation of *exactly* 0) between fatigue and accidents? If we did imagine such a thing, would we be spending time and money investigating this relationship? Does rejecting this very sharply defined hypothesis (i.e. finding *some* relationship between the two factors) really take us very far?

Second, P (D/H), *the probability of the data given an (unlikely) hypothesis*, is not really what we wish to test, which is *P (H/D), the probability of (our particular alternative) hypothesis given the data* that we observe. There is a difference between statistical and *scientific* inference. It is often the case that, with a frequentist approach, statistical and scientific inference may become confused (Sawyer and Peter 1983). For example, if we reject the null hypothesis of no relationship between fatigue and accidents – i.e. find a low enough P (D/H) – we might falsely assume that this means there is a high probability that our particular alternative hypothesis (whatever that might be) about fatigue and accidents is true – i.e. there is a high P (H/D). However, it is not this latter hypothesis that has been tested!

By contrast, the Bayesian approach involves examining precisely what we want to test, i.e. the probability of a model (for example, a theory we have about fatigue and accidents) given the data. This seems to us to be the rightful business of scientists and investigators, but, crucially, demands that we make explicit *the fact that we have a model to start with*. In Bayesian terms this is known as 'setting prior probabilities'. This of course allows different people to have different, 'subjective' estimates of probability because they may be in possession of different information, or have had different experiences. But we have argued in Chapter two that it is unscientific to assume otherwise, and described in Chapter six how presupposition must always be part of the interpretative process.

It has been said elsewhere that a frequentist uses impeccable logic to answer the wrong question, while a Bayesian answers the right question but makes assumptions that nobody can fully believe in. However, in Chapters six and eight we have outlined an approach which we hope avoids viewing subjective judgement as purely relative. If we can determine (via consensus trials) which are our reliable interpretations of existing data (i.e. ones we can 'believe in'), then we can use these to set prior probabilities. These can then be revisited as data accumulate to the advancement of our understanding.

Taxonomies

Taxonomies have two important features, namely their specific content and their general structure or 'geography'. Both these aspects have a bearing on the extent to which a taxonomy is *intrinsically capable* of being used reliably; some systems are difficult or impossible to use reliably, no matter how competent the coders using the system may be. The problem with some widely used taxonomies is that they have never been designed from the bottom up with either of these two features in mind, but like Topsy have simply 'growed' in response to things that happen.

In terms of specific content, regular problems are encountered in terms of categories that are not exclusive; that is in which a particular code is implied or directly subsumed by another, so that it becomes more or less arbitrary which to tick or whether to tick both. For example, a taxonomy might have a code for 'corrosion', and another code for 'corrosion due to poor chemistry'; or a code for 'procedural violation' and another for 'work practice not followed'. In the absence of structure, the latter might be further compounded by a code 'unintentional rule violation', thereby creating a problem not merely of which box to tick, but raising the issue of personal motivation in one category which is absent from the others, and creating a general dilemma of where to put a deliberate rule violation, perhaps performed for a well-intentioned reason. Whatever the basis on which the coder makes the choice, it must come from individual bias/preference since no unambiguous external logic exists to guide the choice.

Prior planning in terms of content, structure and organisation of a taxonomy is essential; whilst it is important that any scheme has the flexibility for the addition of new codes, the discovery that with great regularity one has to add codes because things are happening that are regularly not recognisable in the taxonomy is a sure sign that the taxonomy is inadequate. Taxonomies cannot simply be allowed to grow piecemeal if they are to be used in a consistent way. In general there are two ground rules which in our experience have proved helpful in devising taxonomies, namely all codes must be mutually exclusive *or* where this is not or cannot be the case, hierarchically organised. To the extent that either of these two rules is breached, a taxonomy is intrinsically unreliable in application, regardless of the talents and skills of the coders.

Human error, strategic decision or adaptive action?

In Chapters nine and ten we looked in slightly more detail at some of the less obvious consequences of viewing human action from a cognitivist perspective, and particularly at the groundbreaking work of Rasmussen, even though at the end of the day we tended towards a more connectionist model of mental activity rather than the dominant 'representationalist' view of human cognition. The idea of cognition as a set of 'entities' which in some sense exist in the brain (though in what precise sense is often a somewhat blurred issue), whereby the world is represented in terms of some morphologically analogous internal structure, has much in common with the earlier discredited idea of 'pictures in the brain', so earnestly repudiated by the early behaviourists. We take the view that there are no pictures in the brain; there are no boxes linked by arrows either. It has been argued elsewhere that such pictorial representations of cognition represent the external logical structure of the ideas proposed by the modeller, rather than literally mapping any internal structures in the brain (Davies 1998: 267).

In discussing the much-cited work of Rasmussen, we made two simple points. First, the skill/rule/knowledge-based taxonomy of human error, whereby 'skill-based' refers to failure of skill, 'rule-based' to incorrect or inappropriate interpretation of rules, and 'knowledge-based' to lack of knowledge, bears no relationship whatsoever to the original cognitive theory proposed. To be blunt, the simplicity and apparent utility of the actual words used in Rasmussen's theory has so beguiled most human factors practitioners that an entirely different application has been devised for them than that originally intended.

However, we have also argued not merely against this basic abuse of the original idea, but against cognitivist models in general, even if they are correctly interpreted. Our position is much more akin to saying that the brain has certain properties (like any other object) and that these properties are modified as a result of what happens to it.[2] Similarly, we reject the idea that because a system has certain outputs, these outputs must also have a prior existence within the system. The connectionist and neural-net approaches, whereby a structure actively adapts to and is altered by experience, seem to encapsulate these ideas rather better than approaches based on representation in any meaningful sense.

From the above arguments we have taken a rather large inferential step. If we do not accept traditional models of cognitive representation, then human error due to 'cognitive failure' is no longer possible in those terms. This is especially important where things such as 'failures of attention' are concerned, since as noted in Chapter two the notion of 'attention' has the convenient attributional property that it can be explained in either active ('I wasn't paying attention') or passive ('My attention wandered') terms (note that the whole issue of attributional bias in reports is described in detail in Chapter seven). Thus, when a driver fails to stop at a red light, for example, in some very real sense we can if we wish blame him/her for not managing his/her cognitive processes properly. On the other hand, the language used to describe the sometimes flawed actions of more influential partners in an organisational structure usually centres around their pressing need to make 'strategic decisions', which all sounds rather grand, rather than their potential for not paying attention and mismanaging their cognitions. We have tried to redress this balance by showing that drivers of trains actively and purposively make their own strategic decisions when confronted by the conditions that managers, directors and brutal economics impose upon them. Instead of seeing them as passive and fallible processors of information, who draw on flawed cognitive representations of what the job involves, we feel that greater safety requires an appreciation of the active ways in which they construct and attempt to shape the environments provided for them, and the ways in which 'the brain machine', its properties and potentials, is altered by the circumstances under which it is expected to work. In other words, a different conception of cognition leads to an approach based on adaptation to circumstances, rather than error. On the other hand, such an approach, to be consistent, must apply at all levels of an organisation. For example, things sometimes go wrong at a management level, we suspect, because a manager 'SPADs' his paperwork ('I never received your memo' is probably equivalent to 'I never saw the signal') for reasons very similar to those of a driver who 'SPADs' a signal. Similarly, during a period of inactivity, he may also resort to similar adaptations in order to maintain arousal.

It makes economic sense

The economic issue concerns the practical utility and safety benefits of the systems proposed, compared to the cost of running such a system. Given that new skills are involved, and that time has to be spent analysing reports with a little more rigour and in slightly different ways than is perhaps usually the case, there is a reasonable expectation that the additional effort will be well spent. The argument here has two main components: first the increasing costs of technical fixes relative to the projected costs (human and financial) of the incidents avoided and second the growing subtlety of human error in increasingly technologically sophisticated environments as the more obvious causes of major disasters are progressively eliminated.

Safety costs money. Total safety is therefore infinitely expensive. As a society, we have to set a preferred level of safety for any activity, somewhere between certain death and total safety (if such were possible). This brings us back again to the risk thermostat idea discussed in earlier chapters, but at a societal level it appears less a matter of individual choice or preference and more an uneasy compromise between what level of safety an industry feels is affordable, and what level of risk the public is prepared to tolerate (as distinct from what level they *prefer*) taking into account the monetary cost to them of the service provided and the perceived need for that service.

Whatever one thinks of the ethics of placing a monetary value on human life, the fact of the matter is that public policy and commercial decisions of all kinds are based on the assumption of a common metric and that metric is money. Distasteful though it may seem to some, any sort of planning requires projection of costs of all elements of a system, and people are part of those systems. Given the way the world as a whole operates, and has operated since the earliest days of token economies (as opposed to barter systems), it is difficult to imagine what other measure might by used without starting again and re-inventing the global social/economic world.

Given that this is the way we do things, there are cash limits on safety expenditure; at some point, the monetary value of a human life saved is less than the money necessary to save it. This is a logical, if somewhat indigestible, fact of life in the societies we have constructed for ourselves. Obviously, there are other reasons for saving lives than mere economic viability, but where businesses and profit are concerned these are usually relegated to a position of lesser prominence. In such a situation, it makes sense to take steps to avoid those circumstances that arise when, in the wake of an incident or disaster, the public and political reaction is such that levels of expenditure are demanded for safety fixes that make no economic sense and may not even solve the problem in quite the way that the public and politicians believe – and where such funds might have been spent elsewhere to better effect, if less theatrically. The public and media demand for Automatic Train Protection on the U.K. railways may be such a case in point, involving levels of expenditure per probable fatality avoided that are simply uneconomic, whilst still not filling all the 'holes' in the system that permit SPADs.

It seems reasonable to suggest that as safety improves by equal increments (measured in terms of severity and frequency of incidents for example), the cost of each successive increment increases logarithmically rather than in equal steps, both as a consequence of increasing inflation rates over time, but also as a consequence of the increasing subtlety and complexity of the factors involved as the more obvious 'causes' are successively tackled and removed. What we are suggesting is that, in simple words, the safer something is, then on an average basis the harder, more expensive and less cost-effective it is to make further safety improvements. Media and public reaction when a major accident occurs can also provoke costly responses that would not be justified in a less emotionally charged climate. If such is the case, then prevention of what can reasonably be prevented is cost effective if it impacts upon this apparent 'law of diminishing returns'. A system for early detection and prevention, however, needs to tackle this problem of the escalating costs of one–off technological fixes; and the point about safety reporting is that, correctly done, it reveals things about the motives, intentions, and other aspects of human performance of the people actually involved in doing the work that are not accessible through any channel other than natural language. Finding out why somebody did something, and the assumptions behind the 'why', is often a key feature in understanding the act they actually performed. People do things for reasons, and examination of those reasons often gives clues as to how a problem can be tackled in terms of organisational systems, the main obstacle to which is (usually) resistance to change, rather than redesigning or rebuilding the plant, the main obstacle to which is usually cost.

Encouraging, collecting and analysing safety reports in the ways suggested also costs money, but nothing like as much as buying back public forgiveness after the latest disaster. It seems to us that the relatively small, but continuous, investment necessary to run such systems represents a better deal for both companies, shareholders and the public at large than the massive outlay on new technology that is often necessary to buy back public confidence after things have gone seriously wrong.

The second point is probably less contentious. Major technological advances in safety have been made in all industries over the years; but by comparison the treatment of human factors at the grass roots level has remained relatively underdeveloped until the last two decades, at roughly which point serious commercial interest in the topic began to emerge probably due to the groundbreaking work of Rasmussen in the 1980s, and Reason in the 1990s. Even now, however, we still encounter databases where a plethora (sometimes hundreds) of engineering and technical failures are listed in the greatest detail, along with a single box labelled 'human factors', as though technological issues and human factors were completely separate and independent; as though technological contexts did not actually create the circumstances under which human error occurs and the shape it takes; and as if attaching the words 'human error' to some event was in some way an actual explanation of what went wrong.

Saying that something happened 'due to human error' is not an explanation; it is the start of a process of investigation, not a conclusion except in the broadest sense. It gives no hint as to how to prevent a recurrence. Saying that the *Herald of Free Enterprise* disaster was 'due to human error' has no more precise explanatory value then applying the term 'mechanical failure' to 'explain' the space shuttle *Challenger* disaster. Respectively, one term means little more than 'somebody screwed up' and the other 'a machine broke', neither of which is particularly helpful. But whereas an engineer will want chapter and verse on what component failed, what it was made of, how long it had been in use, and why it snapped, a similarly exploratory approach is seldom adopted to human factors problems, where the label is all too often taken to be the explanation.

Our conclusion is that the time has now come for industries and companies in the 'risk' business to make their decisions on the basis of all the data available to them, and obtainable by them, as problems become successively more multi-factorial, more subtle, and more crucially dependent on the reasons and perceived causes of human action situated in specific technological environments. Simultaneously, there is need to recognise that the time spent on a more painstaking analysis of what people actually say and report is well spent. It merely complements the time and energy spent searching for the causes of technological failures. It accepts that the information in natural reports is a key that can reveal sources of error not available through other means, and that naturalistic text is a source of data equal to any other, but requiring its own methods and skills in interpretation.

Science: induction versus intuition

The basic message of this book is that data are actually or potentially available to high-consequence companies that can enhance safety performance, but that such data are seldom utilised to the full. A reliance on the 'objectivity' of engineering and technical data has led to the relegation of 'subjective' discursive accounts to a position of lesser prominence.[3] It is argued here that the expense of increasing safety through more and more technical fixes is in danger of becoming increasingly cost-ineffective due to ever-increasing costs of improving safety via that route.

It has been argued many times elsewhere that human factors are, by comparison, little understood and often ignored and that message does not need repeating. However, we have argued here that a major source of human factors information is the natural accounts that people provide of what happened, what they did and why. The perceived properties of the circumstances surrounding actions, the interpretation of those circumstances by those involved, and the reasons and motives that lay behind the actions they took, are not accessible through any route other than natural speech. However, methods for reliably turning safety discourses into hard data that can be integrated with, and complement, other data, have seldom been spelled out, either in terms of philosophy or practical implication. Furthermore, the fixes derived from such analysis can often be cheaper but just as effective as technological fixes; furthermore, every piece of technology by its nature provides a context for new sub-species of human error. Technological fixes transform, but do not eliminate, human error.

We are arguing, therefore, in favour of a complementary relationship between technical data and discursive data; the philosophical link that makes such a relationship possible is the suggestion that the distinction between 'hard' and 'soft' data is far from clear-cut and perhaps does not exist at all. All that does exist is data about supposed 'facts', communicated on a largely second-hand basis between individuals, and about which there is or is not agreement. After all, the thing that turns individual technical propositions into 'facts' is that most people agree about their meaning and significance. The same is true, we argue, for the meaning of social facts as presented in safety reports. Given then that the distinction is ideological rather than real, there is a basis for unifying these two types of data.

In a sense, therefore, we have argued against the accepted wisdom that still sees science as basically an inductive process. From such a viewpoint, theory only arises from bodies of objective facts revealed as a consequence of repeated and meticulous observations by motiveless scientists, who speak as the impartial mouthpieces for those 'facts'. Whilst inductivism still enjoys great prestige in some circles, the whole notion is based on outdated philosophies of both science and knowledge. Thus, Gould in his classic work on evolution writes (1982) 'The criticisms of inductivism are certainly valid, and I welcome its dethronement during the last thirty years *as a necessary prelude to better understanding*' (60) (*emphasis added*). Perhaps the only comment to make is that the supposed 'dethronement' still has a little way to go in some circles.

On the other hand, we have been anxious to avoid the bleak and extreme opposite view; that knowledge has no logical basis at all, and depends by contrast on the whims, suppositions and intuitions of individuals. This is the supposed 'eureka' theory whereby reasoned progress becomes impossible, and things happen in an unprincipled and unpredictable way more or less as lightning strikes. In effect, this is basically an anarchical model which elevates human intuition to a level where any sort of reasoned or consensual process underlying the march of the human race through time becomes impossible. Whilst the word 'progress' is sometimes rather difficult to match to the actions that humans have taken from time to time, we do not believe the process is simply chaotic.

What we propose, therefore, is a middle road based on an epistemology that sees agreement as the unifying factor that draws out meaning from data of all types. Whilst the brilliance of Newton in deriving meaning, and thereby practical applications, from the movements of physical objects is indisputable, we suggest there are other phenomena, especially those in which living, thinking people are involved, which require a different basis in science. The things people say about their actions reveal reasons why those actions take place; physical phenomena, whilst they may be said to have causes (but see Chapter seven), can never have *reasons*. The system we have suggested thus makes those things that we see as the causes of events compatible with the *reasons* why people perform actions, bringing them together in a common data form for analysis. Taken together, we see this as an attempt to bring together the whole picture of why things happen, in place of a one-sided picture based only on technical mechanisms. In our view, the time is right both philosophically and economically for such a synthesis in the interests of managing and improving safety and avoiding the increasingly massive costs that can accompany failure to do so.

It is true that the system suggested requires a little more thought than the idiosyncratic inspection and interpretation of what someone said or wrote, by a single individual with actual or assumed expertise. Whilst it is by no means complicated, some knowledge of the nature of language seems desirable; a method of assessing consensus agreement is also necessary, and it may be advisable to carry out one or two simple arithmetic calculations from time to time. Consequently, these are abilities that would be required of anyone using the methods proposed. But such abilities are, and always have been, an element of scientific methods of whatever kind, and surely this is not a major investment compared to what stands to be lost or gained. In the past, if people had not been prepared to listen to what others said, have an ear for things about which there was general agreement, and jot a few symbols and numbers down on the back of an envelope, it is doubtful if we would ever have developed the technologies which are the main concern of this book. It is hardly surprising therefore, if some of the wondrous subtlety underlying modern technologies is ultimately reflected in some small way in the methods which we now believe are necessary for their effective, economic and safe management.

Notes

Chapter 1

1 We exclude from this domain certain specialised and esoteric services, such as 'Mountain Madness' and other extreme adventure companies offering trips to the top of Mount Everest, for which people willingly sign up, despite the fact that one in every eight clients undertaking this adventure fails to come back alive. We assume that this type of service provision is one with a limited appeal, and would not serve as an appropriate model for more general public service provision.

Chapter 2

1 An amusing anecdote was provided by a manager in one plant, where a system of this type had been implemented: 'After six months, we've finally got them to stop. Now we're trying to get them to think.'

2 The word 'biased' in this text is used in a specific sense. It describes the fact that any interpretation takes place from a particular standpoint, which involves the beliefs, attitudes and suppositions of the person making the interpretation. Bias is inherent in the process; by definition there is no such thing as an unbiased opinion. It does *not* imply deliberate falsification. It does *not* imply a source of distortion or error deliberately introduced into the debate. 'Bias' does not mean 'telling lies'.

3 For example, two theories are available concerning the loss of flight TWA 800. One theory suggests that a spark ignited a fuel/air mixture in a central fuel tank caused by break-down of insulation of high-voltage cabling which passed through the tank. The other, largely discredited theory, is that the plane was struck by a rogue missile fired as part of U.S. Naval exercises taking place in the area. Evidence has been put forward in support of both theories. On the basis of the stronger theory, the routing of cables was redesigned, and many existing 747s had their cabling refitted. It is logical, if mischievous, to point out that neither theory can actually be *proved* conclusively, since 'best evidence' is never 'proof'. We *can* say, however, that the modified 747s are safer than the unmodified ones, since a particular type of event can no longer happen, regardless of whether it *ever* happened. If an event prompts a safety improvement, this is a good thing even if the event and the improvement are unrelated, insofar as it stops something else from going wrong! If we accept that we cannot reasonably claim that our causal theories are right every time, then this type of thing must happen. An apocryphal story concerns a pilot's entry in a repair book that supposedly stated 'Something rattling in cockpit'; to which the service engineer replied 'Something tightened up in cockpit'.

4   To be precise, truth is our consensus beliefs about what works.

5   It is interesting to note that the forced analogies we are making between physical and behavioural sciences are not so very forced at this point. Behavioural scientists, for example, regularly make probabilistic statements in terms of significance tests, indicating that so many people will do this, and so many will do that, without offering any help at all with the question as to who precisely will do what.

6   Readers must excuse this use of the word 'know' in connection with inanimate matter. The author merely follows the example of the physicists, who freely speak of sub-atomic particles knowing things and being aware of things. It is by no means clear, from the theory of quantum mechanics, that this is merely a metaphor!

7   A hypothetical example: a train driver stops at a station and a traction inspector notes that he has hung a miniature (toy) pair of football boots on the power take-up lever. The incident is reported, the driver is disciplined. There are no other consequences. Subsequently, in conversation with a researcher, the driver points out that he was stopped at the station facing a red light for over five minutes. He says that when the guard rings the bell there is a tendency to respond to the bell and forget about the red light. 'Ding-ding-and-away' is the phrase he uses; a perfectly adequate way of describing a conditioned response. He puts the toy boots on the lever to remind him to check the colour light before he takes power. The act of putting the toy boots on the lever points to a real problem (the natural tendency to move off when hearing the 'start' signal from the bell) and indicates the need for an engineered reminder device. The data show that taking off from a station against a red light is a not uncommon occurrence. A trivial event thus provides clear clues as to how a future disaster could happen, and how it might be avoided.

8   Even if the cause is a change of mind on someone's part, we have to have information about the change of mind and that would involve communication in some form.

9   If there is, it is of no consequence from the point of view of safety management, since we can't do anything about it.

10 Generally speaking, psychologists have shown that sometimes intentions (from an empiricist position, these are simply acts of speech, not entities in the brain) predict behaviour and sometimes they don't. This somewhat unhelpful conclusion is towed to safety by the 'finding' that intentions are mediated by other variables of equally uncertain epistemological stature, such as beliefs, norms and expectancies.

11 It is worth mentioning studies by de Groot (1978) with expert and non-expert chess players. When presented with random arrangements of pieces, experts and non-experts had an equal (rather poor) capacity subsequently to remember the positions of the pieces. However, when presented with positions from actual games the experts were immediately superior. In periods as short as five seconds, experts could remember general layouts and positions and plan a sensible next move. The ability appeared to have everything to do with recognising general states of affairs on the basis of past experience (i.e. pattern recognition) rather than merely a superior ability to remember the positions of individual pieces. That is, the experts could recognise meaningful, as opposed to meaningless, patterns on the basis of prior experience. In a similar way, we would argue that avoiding disasters may sometimes require an ability to recognise overall configurations of circumstances that are dangerous rather than specific elements. (NB This is not the same as Recognition Primed Decision [RPD] which is more akin to a rule-following algorithm. See Flin [1996] for a discussion of RPD.)

12 We are indebted to Dr Jim Baxter, Dept of Psychology, University of Strathclyde, for coining this phrase.

Chapter 3

1 In statistical terms, there are problems of interpretation when looking for similarities using tests for the significance of difference. In this particular instance, the hypothesis is problematic, since no form of analysis exists to test that things are the *same*. We specifically conclude from this form of analysis that there are no significant differences between the way most of the codes are distributed over the severity levels. We do *not* however claim that we have thereby demonstrated that they are the same.
2 It should be noted that the unit of analysis is not the number of incidents at each level of severity, but the number of causal codes assigned to each level of severity. Full details of this, and all other analyses, are available in Wright (2002).

Chapter 5

1 This is of course an example of the validatory triangulation that Flick (1998) argues against. But the orthodox position is to accept this as long as the 'testing' is numbers over words, not vice versa.
2 The 'picture' is presented without a positive or negative on the Y axis, to avoid having to replicate it 'upside down' for cases where 'too low' is of interest rather than 'too high'.
3 Overall SPAD figures for the company show a downward trend.

Chapter 6

1 A problem because you can't just phone God up and ask him what he meant by a particular passage.
2 And of course this is particularly important when these 'facts' are of cognitive 'inner states'.
3 Interestingly, Rescher sees links between the coherence theory and another major theory of truth, the pragmatic (which states that truth is what is useful). We might suggest that future research should investigate these links and see whether a unified theory is possible). In terms of our own social understanding of the phrase 'legal process' we have simply understood this to mean discussion and debate amongst people attempting to interpret the text (Rescher 1973).
4 They were (as presented on the coding sheet), General Descriptor, Personnel codes, Public codes, Technical, Human Factors, Feedback, Responsibility, Avoidance Action, Environment, Consequences. With the exception of the General Descriptor and Personnel, there was no category which *necessarily* had to be used. The 'environment' section dealt with 'embodied' problems (bad working conditions, heat, cold, etc.), normally omitted from 'cognitivist' coding taxonomies.
5 For the purpose of analysis these codes were sometimes further subdivided, so, for example, rule violations were either single or common. Therefore there were 32 human factors codes in total.

Chapter 7

1 'Causes' cannot be measured, captured in a test tube or analysed under a microscope, so we cannot verify a statement such as 'A was 80 per cent of the cause, and B was 20 per cent'. Causality (especially in complex events) is established once only 'after-the-fact': we cannot *experiment* by 'running' events such as Kegworth or Zeebrugge again, omitting individual elements, to establish what proportion of the 'total causality' individual events account for.

2 This overview of the general attribution area is necessarily brief due to the specific nature of this chapter (for a more detailed account see Antaki 1988a; Hewstone 1983; Kelley and Michela 1980).

3 It should be noted that Neisser (1976) later came to question the usefulness of 'hypothetical models of the mind' (8). As he stated 'If cognitive psychology commits itself too thoroughly to this model, there may be trouble ahead. Lacking in ecological validity, indifferent to culture, even missing some of the main features of perception and memory as they occur in ordinary life, such a psychology could become a narrow and uninteresting specialized field' (1976: 7). Skinner (1984), similarly accuses cognitive scientists of eroding standards of logical thinking and definition and lapsing into a world of speculation.

4 Critiques of cognitivism are rare in the safety management literature. However, as Hollnagel (2002) puts it, describing features of cognition 'often implies assumptions about the nature of cognition that are ambiguous, incorrect or unverifiable' (reference from http://www.ida.liu.se/~eriho/Main_Cognitive Models.html).

5 As noted earlier, work like this can also be cited as evidence for the futility of cognitive approaches. Davies (1997) reports that the Nisbett and Wilson paper was criticised, perhaps because 'it threatened the very basis of a number of social research methods geared to the notion that people's verbal utterances ... described something going on inside their heads' (42–3).

6 Note, once again criticism is directed simultaneously at the idea that language reflects cognitive events and at the idea that it reflects external reality.

7 It is interesting that in the Air Commerce Act of 1926 the department intended to make public the 'causes' of accidents. By 1934, an amendment to the Act determined to make public the 'probable cause or causes' of the accident. By 1938 the safety board of the Civil Aviation Authority were required to report the 'facts, conditions and circumstances relating to each accident'. Note the increasing uncertainty (or *fuzziness*: Kosko 1994) and lack of a deterministic philosophy in the definitions.

8 These authors used the Kappa coefficient to take 'into account the number of agreements that would be expected solely by chance' (Lehane and Stubbs 2001: 121). A discussion of this use of Kappa takes place in Chapter eight.

9 Note that the p value merely denotes the (low) probability of these data if there is no relationship between fatigue and this style of explanation (see Chapter eleven for a discussion).

10 Many have debated whether causal and teleological explanations should be treated differently in attributional analysis, however we concur with those who see no practical benefit in making such a distinction (e.g. Kidd and Amabile 1981).

11 It must of course be noted that some researchers still presume that attributional language can be viewed as a 'window' with which to view cognitive entities (e.g. Silvester *et al.* 1999), and thus try and reconcile attributional construction with the assumption that language reflects mental states. However, Davies (1997) points out that this involves taking 'bits from both these philosophies' despite their 'basic incompatibility' (Davies 1997: 43). Davies postulates that

this is due to a difficulty in treating verbal behaviour as one would treat any other type of behaviour.

12 It has previously been argued (notably by Groeneweg 1992; Hollnagel 1998) that mechanistic (or 'stimulus-response') models of human behaviour are inadequate in a human factors context. A common tactic has been to view *cognitive* events as 'mediating' between stimulus and response (e.g. Rasmussen 1986; Reason 1990). The Cognitive Systems Engineering (CSE) approach (e.g. Woods *et al.* 1994) moves away from the 'information processing' metaphor (see Chapter nine), and involves certain encouraging aspects, namely, that of active processes shaped by human goals and contextual factors. Nevertheless, Hollnagel (1998), in endorsing CSE, still argues that 'human cognition' be 'included in the set of assumed causes' (73). It can be argued that shifting the cause (stimulus) from the environment to the inside of the operator's head does not fundamentally alter the deterministic aspects of the model. In Chapter ten it is proposed that what is needed is a completely different view of the world (a teleological, systems approach, e.g. Benner 1975; Flach 1999) which does not rely on 'chains' of causality, whether they involve cognition or not.

13 This does not allow for people observing a meaningless, external environment. Rather people are part of the environment. This is known as *Mutualism*. As Arthur Still (1998) puts it 'what has evolved is the form of life as a system, which includes both individuals and environment in mutual interdependence' (86). So the attributor's position in the system (i.e. their own function) is closely related to the functional causal language they use.

Chapter 8

1 Alternatively, for example in our own work with minor event coding in the nuclear industry (Wallace *et al.* 2002) there is no requirement to 'choose' a code from a list. Rather each factor is classed as 'present' or 'absent' in a given report.

2 These have been described respectively as the 'external mode of malfunction' (Rasmussen *et al.* 1981) or 'external error mode' (Isaac *et al.* 2000); the 'general failure type' (Reason 1990) or 'external performance shaping factor' (Rasmussen *et al.* 1981); and the 'psychological failure mechanism' (Hollnagel 1998) or 'internal performance shaping factor' (Rasmussen *et al.* 1981).

3 This trend is clearly visible in the essays edited by Wilpert and Qvale (1993). In this book it can be seen that the organisational (see Reason on latent pathogens, Wilpert and Klumb on system safety) and the cognitive (Wagenaar on generic error modelling, Wehner on cognitive structure) go hand in hand in modern human error and safety work. (As will be apparent on reading Chapters nine and ten, we welcome the examination of events from a systems perspective but find less benefit in applying cognitivist and, more generally, 'information processing' principles.)

4 Interestingly, Swain (1990) is quite clear as to the reasons for the adoption of such an approach to human error investigation – the 'hard' technical appearance of the output data has a face validity that is useful in lobbying designers. To be fair, the goal of an integrated system for human and machine reliabilities is best served by a common definition (Leplat and De Terssac 1990), however difficult this may be to justify mathematically (Adams 1982).

5 In fact there are assumptions in Probabilistic Risk Assessment (PRA), on which HRA is modelled, whereby technical items must have single invariate functions and be independent of each other. These assumptions have been shown to be questionable for technical items (Perrow 1984), and especially problematic with

human beings (Timpe 1993; Wickens and Hollands 2000: 501). The principal response to concerns that HRA was too mechanistic has been to 'cognitivise' the HRA arena (e.g. Reason 1990; Hollnagel 1998). However, human reliability is still derived from comparison with some objective benchmark.

6   For completeness, within-rater consensus (WRC) (a measure of a coder's ability to assign the same codes to the same events on separate occasions) might be used to apply to what is usually termed intra-rater reliability. The same main issues would apply to this concept, for example, the fact that consistency in total frequencies or patterns of classification does not necessarily equate to agreement on individual cases.

7   These authors go on to state that the 'TRIPOD' method is distinguished from the other techniques cited because 'it is reliable, i.e. ... different analyses will produce highly similar analyses' (Wagenaar and van der Schrier 1997: 31). However, we have outlined below how 'reliability' as described for TRIPOD does not negate the possibility of disagreement between coders.

8   In fact, Cohen (1960: 43) outlines a case where, under certain conditions, the coefficient of *agreement* Kappa is equivalent to the product-moment *correlation* (ø) for the dichotomous case. However we could find no cases in the literature where ø has been used in this way.

9   Interestingly, a significant difference whereby 'correct' rejections increased over time was shown. The value of an 'objectivist' approach is at once demonstrated, with poor test-retest reliability being interpreted as 'a good thing' (i.e. 'I couldn't agree with myself but I was wrong the last time').

10   It can also be seen that in this case an 'objectivist' approach was employed whereby results from the technique were tested against the 'actual' errors observed. This appears logical enough, if the reliability of a technical item or person involves comparing what happens with what should, then the reliability of a predictive device should be established by comparison with a similar 'benchmark', i.e. what *actually* happens. So the reliability of the technique can be conceptualised as *1 – the difference from the actual error data*. However, in this case the 'actual' data are observational. One of the areas where agreement between observers is essential is observational analysis. So in this study the comparison with 'actual' data is underpinned by observational data for which no reliability data are presented. In the absence of such data, the 'actual' errors may be no more reliable than the data they are being used to test.

11   It is assumed that occasions where no one identified a statement go unnoticed and cannot be part of the analysis.

12   '2 out of 3' and '1 out of 3' assignments have the same average agreement across three pairs of coders at 33%. This is of course because there is no such thing as *1 person agreeing* – in the latter case it is the two coders *who have not picked the code* who agree. In this case, 66% of people said 'no', but this is not the same as reporting this as '66% agreement'. The safest method with multiple raters is to score the mean of all possible paired comparisons as in the examples above and as outlined by Fliess (1971).

13   In addition, trials were carried out on consecutive days, so that 'practice' could be evaluated. Agreement did indeed increase from day 1 (56% for selection and coding; 72% for coding only) to day 2 (66% and 89% respectively). These data show how consensus needs to be continuously evaluated and that 'snapshot' trials can be misleading. Chapter six shows a detailed description of a reliability trial. At the end of the current chapter is a list of recommendations in terms of conditions to be met when trials of this type are carried out.

14   It is interesting to note that seven out of 26 raters were excluded from the trial because they were 'not using the technique appropriately when compared against "benchmark cases"'. It appears this is was actually the first test of

'agreement between raters' but as results were not satisfactory those showing disagreement were merely expelled from the group!

15 Cohen's Kappa (Cohen 1960) is a coefficient which corrects the raw agreement for agreement expected by chance, and is discussed later in this chapter.

16 Feinstein and Cicchetti (1990) call this the 'first paradox of Kappa', stating that 'if Pe (the proportion of agreement due to chance) is large, the correction process can convert a relatively high value of Po (raw agreement) into a relatively low value of K' (544).

17 Strathclyde Event Coding and Analysis System, see Wallace *et al.* 2002

18 Examples of emergent use of Kappa in safety management work include Isaac *et al.* (2000); Lehane and Stubbs (2002).

19 Problems with null hypothesis testing in general have been well documented elsewhere (see, for example, Edwards *et al.* 1963), however, it is beyond the scope of this book to address this wider issue at present (see Chapter eleven).

20 In addition, Kappa is an omnibus index of agreement. It does not allow for distinctions among various types and sources of disagreement. So if an average is computed for various codes, then agreement on individual codes will not be shown. Weighted Kappa (Cohen 1968) allows for the weighting of different disagreements so that the overall Kappa takes these weightings into account. But with ordered category data, one must select weights arbitrarily to calculate weighted Kappa (Maclure and Willet 1987). So once again we would argue that it may be better simply to decide where we are interested in disagreement, and calculate a single coefficient for that particular case. In outlining further criticisms of Kappa, we will assume the coefficient is used to calculate agreement for each code separately.

21 We will not adopt the 'objectivist' term base-rate here, as we are only interested in agreement between two *fallible* sources; however, the problem can be conceptualised in a wider sense by stating that Kappa *varies within set levels of agreement and disagreement depending on marginal totals.*

22 For example, 'human errors' are often cited as accounting for a high percentage of causes assigned (Larson and Merritt 1991), though we have argued (see Chapter seven) that subjectivity and bias in causal investigation can account for such results.

23 Y has also been shown to be problematic in that, like Kappa, it depends 'only on the ratio of observed counts to the total number of observations, and so does not change as the number of observations increases' (Lee and Del Fabbro 2002). Thus 'hard fought' agreement over a large number of trials will be exactly the same as agreement over a handful of trials, provided the *ratio* of 'hits' and 'misses' is the same. In addition, a peculiarity of Y is that the formula gives a perfect agreement of 1 whenever a cell in the matrix = 0. That is, whenever either rater does not assign a particular code at all or always assigns it. Spitznagel and Helzer (1985) propose a 'pseudo-bayes' correction to control for this.

24 A coefficient which is not sensitive to prevalence (RE) was proposed by Maxwell (1977), however RE is still dependent on ratios, and is not sensitive to absolute numbers of observations. Interestingly, Maxwell argues that it is false to argue that raters start from a position whereby 'chance' agreement is a possibility, and so rejected Kappa where agreement is measured against chance.

Chapter 9

1 Note: the theory of mental models was actually developed by Craik in 1943.
However, Craik posited his theory as an *alternative* to the 'rule-following' theory of cognition (although it must be stressed that his theories are entirely
compatible with those of GOFAI) (Craik 1952). However, Rasmussen does not
follow Craik in this respect: Rasmussen's mental models are always representations of external systems states: the 'motor' of cognition is always assumed to be
sequential digital ('rule-following') processing.

2 It might well be questioned how accurate VPA is in terms of gaining access to
'internal cognitive states'. As Chapter seven argues, VPA tends to sideline the
extent to which discourse is functional, not veridical. In a question and answer
session after a presentation, Rasmussen was asked: 'What part of the information processing mechanism is accessible to protocol analysis?' to which
Rasmussen replied: 'I don't know'. He went on to state that it was possible that
perhaps part of the mechanism *might* become possible upon further analysis,
but to the best of our knowledge, no such analysis was done (Rasmussen 1976:
383).

3 Rasmussen's very early work was also influenced by Bruner *et al.* (1956), one of
the 'godfathers of cognitivism'. Bruner's work propounds a view of cognition as
rule bound and symbolic (Sanderson and Harwood 1988).

4 See Gentile 2000; 1972. See also Logan 1988 for a more 'cognitive' alternative
to Fitts and Posner.

5 Now, despite the fact that Simon later claimed that when he discussed the
phrase 'problem representation' he meant 'mental model' (Qin and Simon
1995), this was not generally how his and Simon's work was perceived.
Instead it was seen as being the summation of the algorithmic, 'brain is a digital computer' model of cognition. This is certainly how Rasmussen saw it. He
explicitly states that their over-emphasis on 'process rules' renders their
approach 'not feasible in the present context' (Rasmussen 1980a: 84): his own
model only uses 'process rules' explicitly at the level of RBB although implicitly, as we shall see, at the level of KBB. His use of mental models is not,
therefore, a repudiation of Newell and Simon, merely a statement that 'mental
models' are required *as well as* rules, a view that Simon at least came round to
agreeing with.

6 Moreover, even if it could be 'passed down', how could it be updated? It is clear
that in updating the model (that is, adding new information) the operator
would run into the problem: on what basis was new information decided to be
relevant to updating the model or not? Rules can't be invoked (Rasmussen
states they are not used at this level) Therefore, either *all* the dynamic, infinite
(Toft 1996) amount of information is incorporated into the model (leading to
an infinitely large and complex mental model), or else the operator must build
a separate mental model for each new 'bit' of information coming in (in order
to compare the models in terms of 'usefulness' and accuracy, which might function as a definition of relevance) which, remembering the amount of
information available is infinite, must inevitably lead to an infinite number of
mental models (Heil 1981).
  It is only fair to note that in an attempt to solve this problem Rasmussen has
invoked the thought of the philosopher John Searle and J.J. Gibson and has
argued that it might be possible for update of the mental model at the SBB level
to be via the 'laws of nature – not rules' (Rasmussen 1990: 65). But this misses
the salient point: both Gibson and Searle were *opponents* of 'old style' cognitivism and the 'symbolic' approach (Gibson 1979 and Searle 1990). In fact
Rasmussen admits that Searle's 'background' (which Searle – and Rasmussen –

is using to solve the 'infinite regress' problem) is 'non-representational' (Rasmussen 1990: 45) but describes his (i.e. Rasmussen's) 'dynamic world model' at the level of SBB as being 'a ... *representation* [sic] of objects' (Rasmussen 1990: 55) (*emphasis added*). Rasmussen then claims the 'background' and the 'dynamic world model' are 'basically' the same. However this is clearly incorrect. Searle's and Gibson's ideas are not compatible with Rasmussen's fundamentally cognitivist worldview.

7 This might seem to overstate the problems with 'rule following' but actually it understates them: see Wittgenstein (1953) for even more problems with the 'rule-following' scenario. See also Collins (1990) who points out difficulties resulting from ambiguous rules. See also Eiser (1994) and Ryle (2000).

8 The two terms and the phrase 'neural nets' are not quite synonymous but their meanings are close enough for the purpose of this argument.

9 Moreover, connectionism does seem to be compatible with the Wittgensteinian view of language we have advocated elsewhere in this book (Stern 1991).

10 For example, a plausible 'alternative' to connectionism is Gerald Edelman's 'Neural Darwinism', a theory rooted in Edelman's researches into brain processes. Edelman's theories are useful for demonstrating that cognitivist theories are incompatible with what we know of the physiology of the brain. However, Edelman also criticises connectionism. But are his views really so different from connectionists'? William Clancey notes that attempts to test Edelman's theories take place on connectionist networks. It seems that Edelman's theories develop out of connectionism, rather than contradict them, and that when Edelman argues against connectionist theories, it is the 'first generation' models he has in mind, not twenty-first century dynamic models (Edelman 1994; Clancey 1997).

11 For links between van Gelder's 'dynamicism' and that aspect of systems theory termed 'cybernetics' see van Gelder and Port (1995). For links between dynamicism and Gibson's 'ecological psychology', see Turvey and Carello (1995). See also Noë (2002). For links between Heideggerian hermeneutics and anti-representational thinking see Winograd and Flores (1986).

Chapter 10

1 Therefore this experience is embodied *and* sociated.

2 The key difference of course, (even though Adams mentions arousal), is that the Adams/Wilde theory would predict that when the operator was safe they would seek risk to raise arousal. But of course, arousing behaviour doesn't have to be risky. It's not *risky* to go jogging, watch a movie or video, have a meal, or read a good book, but these can all raise arousal levels. Our theory predicts that we seek information when bored, some of which *can be in the form of* 'risky behaviour'. Therefore we pursue risk because it is an *interesting* state to be in, and interesting things raise arousal. It's obvious that our information arousal paradigm has far broader implications for human behaviour.

3 Therefore we have moved back from qualitative to quantitative analysis. It should be noted that new studies are now being carried out by the train company to test the predictions of IAT. The circular process continues.

4 The equivalent state of 'boredom' leading to 'error' in the study of road vehicle drivers is often termed 'highway hypnosis'. Perhaps 'railway repetition' might be a similarly alliterative way of describing our own findings (Tejero and Chóliz 2002).

5 Some companies do in fact provide such a facility.

6 Variety has a technical definition: the amount of possible states a system may hold in a topological 'state space'. This can be defined mathematically.

However, the 'ordinary language' definition of variety usually functions as a perfectly adequate use of the word in this context.

7   Possibly on the neural equivalent of a 'fuzzy' connectionist network (Kosko 1994).

Chapter 11

1   The use of the word 'tool' is an interesting case in point. Defined by the *Concise Oxford Dictionary* as 'any device or implement used to carry out mechanical functions', this useful and hitherto unambiguous word has been stolen by psychologists and other social scientists to refer to virtually anything they produce; questionnaires, procedures and even diagnostic criteria. Since language is performative, we might speculate that the intention is to give the impression that such things are as straightforwardly and practically useful as hammers and screwdrivers. This is probably not always the case!

2   To give a couple of naïve examples: suppose I crash my car into a lamp-post such that it no longer provides the output of transport. I may represent this process in terms of boxes labelled 'lamp-post', 'crash', and 'transport failure', and link these by arrows. None of these things is 'represented' in the car in any sense that implies that I can *find them in the car if I look for them*, which raises the question as to the utility and epistemological status of such a form of description. The car has simply been altered by its experience, and its properties and potentials after the crash are merely different from what they were prior. With respect to the second issue of whether outputs from a system are represented within it, the best example is the dynamo. The dynamo is an arrangement of wires and magnets, such that if you do certain dynamic things to it, its static properties are altered and it produces electricity. There is no electricity in a dynamo however, and electricity is not 'represented' in any shape or form within a dynamo.

3   For example, a search of a power-plant database revealed a preponderance of technical codes and a virtual absence of meaningful human factors codes. This was partly due to the style of the reports, in which there was an absence of personal pronouns. It remains a mystery why the use of words like 'I', 'we', 'she' 'he' is eschewed in the interests of 'scientific' and exact reportage, and why so many students still get their knuckles rapped for using these words. The phrase 'I collected data from 30 people I bumped into on Sauchiehall Street' is both more honest and more precise than a formula such as 'Data were collected from a random sample in Glasgow', telling us both who collected the data and how the subjects were contacted. By contrast, the 'scientific' version amounts to blatant impression management. 'Economical with the truth' is a good way to describe such a reporting style.

Bibliography

Abramson, L.Y., Garber, J. and Seligman, M.E.P (1980) 'Learned helplessness in humans: an attributional analysis', in J. Graber and M.E.P. Seligman (eds) *Human Helplessness: Theory and Applications* New York: Academic Press.

Adams, J.J. (1982) *Simulator Study of A Pictorial Display for Instrument Flight* (NASA Technical Paper no. 1963) Hampton, VA: NASA Langley Research Centre.

Adams, J. (1995) *Risk*. London: UCL Press.

Adams, N.L. and Hartwell, N.M.(1977) 'Accident reporting systems: a basic problem area in industrial society', *Journal of Occupational Psychology*, 50, 285–98.

Alverman, M. (1999) 'Risky business', in *Salon.com*, 8, 22. Online. Available HTTP: http://www.salon.com/health/feature/1999/07/22/risk/print.html (accessed 11 September 2002).

Antaki, C. (ed.) (1988a) *Analysing everyday explanation: A casebook of methods*. London: Sage.

Antaki, C. (1988b) 'Explanations, communication and social cognition' in Antaki, C. (ed.) *Analysing everyday explanation: A casebook of methods*. London: Sage.

Applied Anthropology (2001) *Lifeguard vigilance: a bibliographic study*. Paris: Applied Anthropology. Online. Available HTTP: http://www.poseidon-tech.com/us/vigilanceStudy (accessed 11 September 2002).

Apter, M.J. (1992) *The Dangerous Edge: The Psychology of Excitement*. New York: The Free Press.

Argyle, M. (1972) *The Social Psychology of Work*. London: Pelican.

Ashby, W.R. (1956) *An Introduction to Cybernetics*. London: Chapman and Hall.

ASRS (2002) (Aviation Safety Reporting System) Online. Available HTTP: http://asrs.arc.nasa.gov/main.htm (accessed 21 October 2002).

Astington, J. (1998) 'What is theoretical about a child's theory of mind? A Vygotskian view of its development', in P. Carruthers and P. Smith (eds) *Theories Of Mind*. Cambridge: Cambridge University Press.

Austin, J.L. (1962a) *How to do Things with Words*. Oxford: Clarendon Press.

Austin, J.L. (1962b) *Sense and Sensibilia*. Oxford: Oxford University Press.

Ayer, A.J. (1936) *Language, Truth and Logic*. London: Gollancz.

Baber, C. and Stanton, N.A. (1996) 'Human error identification techniques applied to public technology: predictions compared with observed use', *Applied Ergonomics*, 27, 2: 119–31.

Bainbridge, L. (1987) 'Increasing levels of automation can increase, rather than decrease, the problems of supporting the human operator', in J. Rasmussen, K. Duncan, and J. Leplat, (eds) *New Technology and Human Error*. Chichester: Wiley. Online. Available HTTP: http://www.bainbrdg.demon.co.uk/Papers/ Ironies.html (accessed 11 September 2002).

Bainbridge, L. (1989) 'The relation between the categories in "types of skill" and in Rasmussen's 'skill-rule-knowledge based' schema'. Unpublished note. Online. Available HTTP: http://www.bainbrdg.demon.co.uk/Papers/RasmusSRKB.html (accessed 11 September 2002).

Bakan, D. (1967) *On Method*. San Francisco: Jossey-Bass.

Barber, B. (1983) *The Logic and Limits of Trust*. New Brunswick: Rutgers University Press.

Barab, S.A., Barnett, M., Yamagata-Lynch, L., Squire, K. and Keating, T. (1999) 'Using activity theory to understand the contradictions characterizing a technology-rich introductory astronomy course', Paper presented at the 1999 Annual Meeting of the American Educational Research Association. Online. Available HTTP: http://www.quasar.ualberta.ca/edpy597/readings/m15_ research.pdf (accessed 11 September 2002).

Bechtel, W. and Abrahamson, A. (2002) *Connectionism and the Mind*. Oxford: Blackwell.

Bell, P.A., Greene, T., Fisher, J., Baum, A. (1996). *Environmental Psychology*. Fort Worth: Harcourt Brace College Publishers.

Benner, L.(1975) 'Accident theory and accident investigation', in *Proceedings: Society of Air Safety Investigators Annual Seminar*, Ottawa, Canada 7–9 October, 1975.

Berlyne, D.E. (1960) *Conflict, Arousal and Curiosity*. London: Mcgraw-Hill.

Berlyne, D.E. (1971). *Aesthetics and Psychobiology*. New York: Appleton-Century-Crofts.

Berlyne, D.E. (1974a). 'The new experimental aesthetics', in Berlyne (ed.) *Studies in the New Experimental Aesthetics*. Washington D.C.: Hemisphere Publishing Corporation.

Berlyne, D.E. (1974b). 'Novelty, complexity and interestingness', in Berlyne (ed.) *Studies in the New Experimental Aesthetics*. Washington, D.C.: Hemisphere Publishing Corporation.

Berman, J. (1996) 'Confidential event reporting (HF/GNSR/5009). Final report. Document reference: 112/C/03/il 1996', *Report submitted under the 1995/1996 IMC GNSR Programme*. Commercial-in-confidence.

Billings, C.E. and Reynard, W.D. (1984) 'Human factors in aircraft incidents: results of a 7-year study,' *Aviation, Space and Environmental Medicine*, 55:960–5.

Bird, F.E. (1966) *Damage Control*. Philadelphia: Insurance Company of North America.

Blalock, H.M. (1964) *Causal Inference in Non-Experimental Research*. Chapel-Hill: University of Carolina.

Blauberg, I.V, Sadovsky, V.N., and Yudin, E.G., (1977) *Systems Theory: Philosophical and Methodological Problems*. Moscow: Progress Publishers.

Bleicher, J. (1980) *Contemporary Hermeneutics*. London: Routledge and Kegan Paul.

Blumenthal, M. (1970) 'An alternative approach to measurement of industrial safety performance based on a structural conception of accident causation,' *Journal of Safety Research*, 2: 123–30.

Bonatti, L. (1994) 'Propositional reasoning by model?', *Psychological Review*, 101, 4, 725–33.

Borg, W. and Gall, M. (1989) *Educational Research*. London: Longman.

Boring, E.G. (1923) 'Intelligence as the tests test it', *New Republic*, 34: 35–7.

Brannen, J. (1992). *Mixing Methods: qualitative and quantitative research*. Aldershot, England: Avebury.

Breakwell, G.M., Hammond, S. and Fife-Schaw, C. (eds) (1995) *Research methods in psychology*. London: Sage.

Brewer, W. (1995). 'To assert that all human knowledge and memory is represented in terms of stories is certainly wrong', in R.S. Wyer (ed.) *Knowledge and Memory: the real story*. Hove: Lawrence Erlbaum.

Brooks, R.A. (1987) 'Intelligence without representation', *Artificial Intelligence*, 47: 139–59. Online. Available HTTP http://www.ai.mit.edu/people/brooks/papers/representation.pdf (accessed 11 September 2002).

Brown, K.A., Willis, P.G. and Prussia, G.E. (2000) 'Predicting safe employee behaviour in the steel industry: development and test of a sociotechnical model', *Journal of Operations Management*, 18: 445–65.

Bruner, J.S., Goodnow, J.J. and Austin, G.A. (1956) *A Study of Thinking*. New York: Wiley.

Campbell, D.T. and Fiske, D.W. (1959) 'Convergent and discriminant validation by the multitrait-multimethod matrix', *Psychological Bulletin*, 56(2): 81–105.

Cardwell, J.D. (1971) *Social Psychology: a Symbolic Interaction Perspective*. Philadelphia: F.A. Davis Company.

Carey, G and Gottesman, I.I. (1978) 'Reliability and validity in binary ratings: areas of common misunderstanding in diagnosis and symptom ratings', *Archives of General Psychiatry*, 35: 1,454–9.

Carlin, B.P., and Louis, T.A. (2000) *Bayes and Empirical Bayes Methods for Data Analysis*. New York: Chapman and Hall.

Caro, T.M., Roper, R., Young, M. and Dank, G.R. (1979) 'Inter-Observer reliability', *Behaviour*. LXIX, 3–4: 303–15.

Carver, C.S. and Scheier, M.F. (1998) *On the Self-Regulation of Behaviour*. Cambridge: Cambridge University Press.

Cassirer, E. (1944) *An Essay on Man*. New Haven: Yale University Press.

CHIRP (2002) (Confidential Human Factors Incident Reporting Programme). Online. Available HTTP: http://www.chirp.co.uk/ (accessed 21 October 2002).

Chouliaraki, L. and Fairclough, N. (2001) *Discourse in Late Modernity: Rethinking Critical Discourse Analysis*. Edinburgh: Edinburgh University Press.

Christoffersen, K., Hunter, C. and Vicente, K. (1996) 'A longitudinal study of the effects of ecological interface design on skill acquisition', *Human Factors*, 38, 3: 523–41.

Cicchetti, D.V and Feinstein, A.R. (1990) 'High agreement but low kappa: II. Resolving the paradoxes', *Journal of Clinical Epidemiology*, 43, 6: 551–8.

Clancey, W. (1993) 'Situated action: A neuropsychological approach', *Cognitive Science*, 17: 98–107. Online. Available HTTP: https://postdoc.arc.nasa.gov/cgibin/postdoc/get?url_id=6662&ext=pdf (accessed 11 September 2002).

Clancey, W.J. (1997) *Situated Cognition: On Human Knowledge and Computer Representations*. Cambridge: Cambridge University Press.

Clarke, S. (1998) 'Safety culture on the UK railway network', *Work and Stress*, 12, 1: 6–16.

Cochrane, R. and Duffy, J. (1974) 'Psychology and scientific method', *Bulletin of the British Psychological Society*, 27, 117–21.

Cohen J. (1960) 'A coefficient of agreement for nominal scales', Educational and Psychological Measurement, 20: 37–46.

Cohen J. (1968) 'Weighted kappa: nominal scale agreement with provision for scaled disagreement or partial credit', *Psychological Bulletin*, 70: 213–20.

Cohen, N. and Stewart, I. (1994) *The Collapse of Chaos*. London: Penguin.

Collins, H.M. (1990) *Artificial Experts*. London: MIT Press.

Costa, G. (1995) 'Occupational stress and stress prevention in air traffic control', Geneva: International Labour Office. Online. Available HTTP: http://www.ilo.org/public/english/protection/condtrav/pdf/1stress.pdf (accessed 11 September 2002).

Craik, K. (1952) *The Nature of Explanation*. London: Cambridge University Press.

Cummings, T.G. and Worley, C.G. (1997) *Organization Development and Change*. Chicago: South Western.

Cziko, G. (2000) *The Things We Do: Using the Lessons of Bernard and Darwin to Understand the What, How, and Why of Our Behaviour*. Massachusetts: MIT Press.

D'Arcy, J.F. (2000) *Air Traffic Control Specialist Decision Making and Strategic Planning – A Field Study*. Atlantic City: U.S. Department of Transportation. Online. Available HTTP: http://www.hf.faa.gov/docs/wjhtc/tn0029.pdf (accessed 11 September 2002).

Davidson, Donald (2001) *Inquiries into Truth and Interpretation*. Oxford: Oxford University Press.

Davies, J.B. (1992) *The Myth of Addiction*. London: Psychology Press.

Davies, J.B. (1996) 'Health research: need for a methodological revolution?', *Health Education Research*, 11, 2: i–iv.

Davies, J.B. (1997) *Drugspeak: the Analysis of Drug Discourse*. Amsterdam: Harwood Academic Publishers.

Davies, J.B. (1998) 'Pharmacology versus social process: competing or complementary views on the nature of addiction?', *Pharmacology and Therapeutics*, 80, 3: 265–75.

Davies, J.B. and Baker, R. (1987) 'The impact of self-presentation and interviewer-bias effects on self-reported heroin use', *British Journal of Addiction*, 82: 907–12.

Davies, J.B. and Best, D. (1996) 'Demand characteristics and research into drug use', *Psychology and Health*, 11: 291–9.

Davies, J.B., Henderson, M. and Hutcheson, G. (1997) *Alcohol in the Workplace: Results of an Empirical Study*. Copenhagen: World Health Organisation.

Davies, J.B., Wright, L., Courtney, E. and Reid, H. (2000) 'Confidential incident reporting on the UK railways: The 'CIRAS' System', *Cognition, Technology and Work*, 2: 117–25.

De Groot, A.D. (1978) *Thought and Choice in Chess*. The Hague: Mouton.

Dejoy, D.M. (1994) 'Managing safety in the workplace: an attributional theory analysis and model', *Journal of Safety Research*, 25: 3–17.

Denison, D. and Spreitzer, G. (1991) 'Organizational culture and organizational development: a competing values approach', in R. Woodman and W. Posmore (eds) *Research in Organisational Change and Development*, Vol. 5. Greenwich: JAI Press.

Dennett, D.C. (1990) 'Cognitive wheels: the frame problem of AI', in M. Boden (ed.) *The Philosophy of Artificial Intelligence*. Oxford: Oxford University Press.

Denzin, N.K. (1978) *The Research Act. A Theoretical Introduction to Sociological Methods*. New York: McGraw Hill.

Devitt, M. (1984) *Realism and Truth*. Princeton, NJ: Princeton University Press.

Dougherty, E.M. (1990) 'Human reliability analysis – where shouldst thou turn?', *Reliability Engineering and Systems Safety*, 29: 283–99.

Doyle, J.K. and Ford, D.N. (1998) 'Mental models concepts for systems dynamics research', *Dymanics Research*, System Dynamics Review, in press. Online. Available HTTP http://www.wpi.edu/Academics/Depts/SSPS/Faculty/Papers/06.pdf (accessed 11 September 2002).

Dreyfus, H. (1994) *What Computers Still Can't Do*. London: MIT Press.

Dreyfus, H. (1998) 'Intelligence without representation'. Unpublished note. Online. Available HTTP http://www.hfac.uh.edu/cogsci/dreyfus.html (accessed 4 April 2002).

Dreyfus H. and Dreyfus S. (1990) 'Making a mind versus modelling the brain: artificial intelligence back at a branch-point,' in M. Boden (ed.) *The Philosophy of Artificial Intelligence*. Oxford: Oxford University Press.

Edelman, G. (1994) *Bright Air, Brilliant Fire*. London: Penguin.

Edkins, G. and Pollock, C. (1997) 'The influence of sustained attention on railway accidents', *Accident Analysis and Prevention*, 29, 4: 533–39.

Edwards, D., Ashmore, M. and Potter, J. (1995) 'Death and furniture: the rhetoric, politics and theology of bottom line arguments against relativism', *History of the Human Sciences*, 8: 25–9.

Edwards, G., Babor, T.F., Raw. M. and Stockwell, T. (1995) 'Playing fair; science, ethics and scientific journals', *Addiction*, 90: 3–8.

Edwards, W., Lindman, H. and Savage, L.J. (1963) 'Bayesian statistical inference for psychological research', *Psychological Review*, 70: 193–242.

Edwards, D. and Potter, J. (1992) *Discursive Psychology*. London: Sage.

Eiser, J.R (1983) 'From attributions to behaviour', in M. Hewstone (ed.) *Attribution Theory: Social and Functional Extensions*. Oxford: Blackwell.

Eiser, J.R., (1994) *Attitudes, Chaos and the Connectionist Mind*. London: Blackwell.

Eiser, J.R., Sutton, S.R. and Wober, M. (1978) '"Consonant" and "dissonant" smokers and the self-attribution of addiction', *Addictive Behaviours*, 3: 99–106.

Embrey, D.E. (1986) 'SHERPA: a systematic human error reduction and prediction approach', *in Proceedings of the International Topical Meeting on Advances in Human Factors in Nuclear Power Systems*, LaGrange Park: American Nuclear Society.

Embrey, D.E., Humphreys, P.C., Rosa, E.A., Kirwan, B. and Rea, K. (1984) *SLIM-MAUD: An Appraisal to Assessing Human Error Probabilities Using Expert Judgement Nureg/Cr-3518, USNRC, Washington DC- 20555*. Brookhaven: Brookhaven National Lab.

Falla, M. (ed.) (1999) *Advances in Safety Critical Systems: Results and Achievements from the DTI/EPSRC R&D Programme in Safety Critical Systems*. London: Department of Trade and Industry. Online. Available HTTP http://www.comp.lancs.ac.uk/computing/resources/scs/index.html (accessed 11 September 2002).

Feinstein, A.R. and Cicchetti, D.V. (1990) 'High agreement but low kappa: the problems of two paradoxes', *Journal of Clinical Epidemiology*, 43, 6: 543–9.

Ferry, T.S. (1988) *Modern Accident Investigation and Analysis*. New York: Wiley and Sons.

Feggetter, A.J. (1982) 'A method for investigating human factor aspects of aircraft accidents and incidents', *Ergonomics*, 25: 1,065–75.

Feynman, R.P. (1985) *QED: the Strange Theory of Light and Matter*. Princeton, NJ: Princeton University Press.

Fielding, N.G. and Fielding, J.L. (1986) *Linking Data (Qualitative Research Methods, Vol.4)*. London: Sage.

Finkel, L. (1990) 'A model of receptive field plasticity and topographic map reorganisation in the somatosensory cortex', in J. Hanson and C. R. Olsen (eds) *Connectionist Modelling and Brain Functioning*, London: MIT Press.

Fischoff, B., Lichtenstein, S., Slovic, P., Derby, S. and Keeney, R. (1981) *Acceptable Risk*. Cambridge: Cambridge University Press.

Fishbein, M., and Ajzen, I. (1975) *Belief, Attitude, Intention, and Behavior: An Introduction to Theory and Research*. Reading, MA: Addison-Wesley.

Fisher, C.D. (1993) 'Boredom at work: a neglected concept', *Human Relations*, 46, 3.

Fitts, P.M., and Posner, M.I. (1967) *Learning and Skilled Performance in Human Performance*. Belmont CA: Brock-Cole.

Flach, J.M. (1995) 'The ecology of human machine systems: a personal history', in J. Flach, P. Hancock, J. Caird, and K. Vicente (eds) *Global Perspectives on the Ecology of Human-Machine Systems*. Hillsdale: Lawrence Erlbaum.

Flach, J.M. (1999) 'Beyond error: The language of co-ordination and stability' in P.A. Hancock (ed.) *Human Performance and Ergonomics*. London: Academic Press.

Fleiss, J.L. (1971) 'Measuring nominal scale agreement among many raters', *Psychological Bulletin*. 76: 378–81.

Fleiss, J.L. (1981) *Statistical Methods for Rates and Proportions*. New York: John Wiley and Sons.

Flick, U. (1992). 'Triangulation revisited: strategy of validation or alternative?', *Journal for the Theory of Social Behaviour*, 22, 2: 175–97.

Flick, U. (1998) *An Introduction to Qualitative Research*. London: Thousand Oaks, New Delhi: Sage.

Flin, R. (1996) *Sitting in the Hot Seat*. Chichester: Wiley.

Fodor, J. and Pylyshyn, Z. (1995) 'Connectionism and cognitive architecture: a critical analysis', in C. Macdonald, and G. Macdonald (eds) *Connectionism: Debates on Psychological Explanation, Volume Two*. Oxford: Blackwell.

Følstad, A. (1999) 'How is a system interface to be designed for optimal support of expert problem solving in complex human-machine systems?', unpublished thesis, Norwegian University of Science and Technology. Online. Available HTTP:http://www.svt.ntnu.no/psy/Bjarne.Fjeldsenden/TermPapers/FolstadInte rface.html (accessed 11 September 2002).

Forsyth, D.R and McMillan, J.H. (1981) 'Attributional affect and expectations: a test of Weiner's three dimensional model', *Journal of Educational Psychology*, 73, (3): 393–403.

Foucault, M. (1972) *The Order Of Things*. London, Tavistock.

Freeman, W. (1997) 'Three centuries of category errors in studies of the neural basis of consciousness and intentionality', *Neural Networks*, 10, 1,175–83. Online. Available HTTP: http://sulcus.berkeley.edu/wjf/AD-.Category.errors.pdf (accessed 11 September 2002).

Freeman, W. and Skarda C. (1990) 'Representations, who needs them?', in J. McGaugh, N. Weinberger, G. Lynch (eds) *Brain Organization and Memory Cells, Systems, and Circuits*. Oxford: Oxford University Press. Online. Available HTTP http://sulcus.berkeley.edu/FreemanWWW/manuscripts/IC10/90.html (accessed September 2002).

Frenk, H. and Dar, R. (2000) *A Critique of Nicotine Addiction*. Boston: Kluwer Academic Publishers.

Frese, M. and van Dyck, C. (1996) 'Error management: Learning from errors and organizational design,' Paper presented at the annual meeting of the Academy of Management, Cincinatti, August.

Gadamer, H.-G. (1981) *Truth and Method*. London: Sheed and Ward.

Gelman, A., Carlin, J.B., Stern, H.S., and Rubin, D.B. (1995) *Bayesian Data Analysis*. London: Chapman and Hall.

Gentile, A.M. (1972) 'A working model of skill acquisition with application to teaching', *Quest*, 17: 3–23.

Gentile, A.M. (2000) 'Skill acquisition: action, movement and neuromotor processes', in J.H. Carr and R.B. Shepherd (eds) *Movement Sciences: Foundation for Physical Therapy in Rehabilitation*. Maryland: Aspen Press.

Gephart, R.P. (2001) 'Ethnostatistics, research methods and organizational behavior' in D. Boje (ed.) *Research Methods Forum 2001*. New York, Research Methods Division, Academy of Management, Briarcliff Manor. Online. Available HTTP http://www.aom.pace.edu/rmd/2001forum/esrmob.pdf (Accessed 11 October 2002).

Geyer, F. and van der Zouwen, J. (1991) 'Cybernetics and social science, theories and research in socio-cybernetics', *Kybernetes*, 20(6): 81–92. Online. Available HTTP http://www.unizar.es/sociocybernetics/chen/pfge3.html (accessed 11 September 2002).

Gibson, J.J. (1979) *The Ecological Approach to Visual Perception*. Boston: Houghton Mifflin.

Gilhooly, Wood, Kinnear and Green (1988) 'Skill in map reading and memory for maps', *The Quarterly Journal Of Experimental Psychology*. 40A: 87–107.

Glass, D.C. and Singer, J.E. (1972) *Urban Stress: Experiments on Noise and Social Stressors*. New York: Academic Press.

Gleick, J. (1998) *Chaos: the Amazing Science of the Unpredictable*. London: Random House.

Gould, S.J. (1982) *The Panda's Thumb*. New York: Norton and Co.

Green, A. and Gilhooly, K. (1992) 'Empirical advances in expertise research', in M. Keane and K. Gilhooly (eds) *Advances in the Psychology of Thinking, Volume One*. London: Harvester Wheatsheaf.

Greene, J. and D'Oliveira, M. (1982) *Learning to Use Statistical Tests in Psychology: A Student's Guide*. Milton Keynes: Open University Press.

Greve, W. (2001a) 'Free will as a problem of psychology; ascribing guilt and the voluntariness of action', Institut für Psychologie, Hildesheim.

Greve, W. (2001b) 'Traps and gaps in action explanation: theoretical problems of a psychology of human action', *Psychological Review*, 108, 2: 435–51.

Gribbin, J. (1995) *Schroedinger's Kittens and the Search for Reality*. London: Weidenfeld and Nicolson.

Groeneweg, J. (1992) *Controlling the Controllable: The Management of Safety*. Leiden: DSWO Press.

Groeneweg, J. (1996, 3rd revised edn) *Controlling the Controllable: The Management of Safety*. Leiden: DSWO Press.

Groeneweg, J. (1998, 4th edn) *Controlling the Controllable: the Management of Safety*. Leiden: DSWO Press.

Grove, W.M., Andreasen, N.C., McDonald-Scott, P., Keller, M.B. and Shapiro, R.W. (1981) 'Reliability studies of psychiatric diagnosis', *Archives of General Psychiatry*, 38: 408–13.

Habermas, J. (1977) 'A review of Gadamer's *Truth and Method*', in F. Dallmayr, and T. McCarthy (eds) *Understanding and Social Enquiry*. London: University of Notre Dame Press.

Hale, A.R. 'Conditions of occurrence of major and minor accidents', Paper given at the *2me Séance Du Séminaire 'Le Risque De Défaillance Et Son Controle Par Les Individus Et Les Organisations'*. Gif sur Yvette. 6–7 Novembre 2000.

Hale, A.R. and Hale, M. (1972) *A Review Of The Industrial Accident Research Literature*, Research paper to the Committee on Safety and Health at Work. London: HMSO.

Hammersley, M. (1992) 'Deconstructing the qualitative-quantitative divide', in J. Brannen (ed.). *Mixing Methods: Qualitative and Quantitative Research*. Aldershot: Avebury.

Harrison, P. I. (1991) 'Harnessing operational experiences and learning lessons: the value of confidential incident reporting schemes,' in *Proceedings of the 11th Annual Safety and Reliability Society Symposium: Offshore Safety And Reliability*. Oxford: Elsevier.

Harvey, J.H., Turnquist, D.C. and Agostineli, G. (1988) 'Identifying attributions in oral and written material', in C. Antaki. (ed.) *Analysing Everyday Explanation: A Casebook of Methods*. London: Sage.

Heidegger, M. (1926/1962) *Being and Time*. London: SCM Press.

Heider, F. (1958) *The Psychology of Interpersonal Relations*. New York: Wiley.

Heider, F. and Simmel, M. (1944) 'An experimental study of apparent behaviour', *American Journal of Psychology*, 57: 243–9.

Heil, J. (1981) 'Does cognitive psychology rest on a mistake?', *Mind*, XC: 321–42.

Heinrich, H.W. (1931) *Industrial Accident Prevention*. New York: McGraw-Hill.

Heinrich, H.W., Petersen, D. and Roos, N. (1980. 5th edn) *Industrial Accident Prevention*. New York: McGraw-Hill.

Hekman, S. (1986) *Hermeneutics and the Sociology of Knowledge*. Oxford: Polity Press.

Hendrick, K. and Benner, L. (1987) *Investigating Accidents with STEP*. New York: Marcel Dekker.

Heritage, J. (1984) *Garfinkel and Ethnomethodology*. Cambridge: Polity.

Hewstone, M. (1983) 'Introductory overview', in M. Hewstone (ed.) *Attribution Theory: Social and Functional Extensions*. Oxford: Blackwell.

Heybroek, R. (1995) *Improving Safety Training: Human Factors Discrepancies Report*. Vosper Thornycroft (UK) Limited, MSC Division. Commercial-in-confidence.

Heylighen, F. (1993) 'Epistemology, introduction' in F. Heylighen, C. Joslyn, and V. Turchin. (eds) *Principia Cybernetica Web*, Brussels: Principia Cybernetica. Online. Available HTTP http://pespmc1.vub.ac.be/EPISTEMI.html (accessed 20 September 2002).

Heylighen, F. and Joslyn, C. (2001) 'Cybernetics and second-order cybernetics', in R.A. Meyers (ed.) *Encyclopedia of Physical Science and Technology*. New York: Hodder and Stoughton. Online. Available HTTP: http://pespmc1.vub.ac.be/Papers/Cybernetics-EPST.pdf (accessed 11 September 2002).

Hidden, A. (1989) *Investigation into the Clapham Junction Railway Accident*. London: The Stationery Office.

Hiroto, D. and Seligman, M. (1975) 'Generality of learned helplessness in man', *Journal Of Personality And Social Psychology*, 31, 2: 311–27.

Hobbs, A. and Williamson, A. (2002) 'Skills, rules and knowledge in aircraft maintenance: errors in context', *Ergonomics*, 45, 4: 290–308.

Hodges, W. (1993) 'The logical content of theories of deduction,' *Behavioral and Brain Sciences*, 16, 2: 353–4.

Hofman, D.A. and Stetzer, A. (1998) 'The role of safety climate and communication in accident interpretation: Implications for learning from negative events', *Academy of Management Journal*, 41, 6: 644–57.

Hollnagel, E. (1983) 'Why human error is a meaningless concept', Paper presented at Nato Conference on Human Error, Bellagio, Italy. Online. Available HTTP http://www.iav.ikp.liu.se/staff/hollnagel/HE%20discussion.htm (accessed 11 September 2002).

Hollnagel, E. (1998) *Cognitive Reliability and Error Analysis Method*. Oxford: Elsevier Science Ltd.

Horgan, T. and Tienson, J. (1996) *Connectionism and the Philosophy of Psychology*. London: MIT Press.

HSE Climate Survey Tool (1998). Online. Available HTTP www.hse.gov.uk (accessed 20 September 2002).

HSE (2000) *Successful Health and Safety Management*. London: HSE Books.

http://www.class.uidaho.edu/mickelsen/ToC/hume%20treatise%20ToC.htm (accessed 20 September 2002).

Hume, D. (1739) *A Treatise on Human Nature*. Online. Available HTTP

Hume, D. (1748/1962) *Enquiry Concerning Human Understanding*. Oxford: Clarendon Press.

Hutchins E. (1983) 'Understanding micronesian navigation', in D. Gentner and A. Stevens (eds) *Mental Models*. London: Lawrence Erlbaum.

Hutchins, E. (1995) *Cognition in the Wild*. Cambridge, MA: MIT Press.

Hutter, B.M. (2001) *Regulation and Risk*. Oxford: Oxford University Press.

INSAG (International Nuclear Safety Advisory Group) (1991) *Safety Culture*. Safety Series No. 75- INSAG-V Vienna: IAEA.

Isaac, A., Shorrock, S.T. and Kirwin, B. (2002) 'Human error in European air traffic management: the HERA project', *Reliability Engineering and System Safety*, 75: 257–72.

Isaac, A., Shorrock, S., Kirwin, B., Kennedy, R., Anderson, H. and Bove, T. (2000) 'Learning from the past to protect the future- the HERA approach', Paper presented at the 24th European Association for Aviation Psychology Conference, Crieff, Scotland, September.

Ives, G. (1991) 'Near miss reporting pitfalls for nuclear plants', in T.W. Van der Schaaf, D.A. Lucas, and A.R. Hale (eds) *Near Miss Reporting as a Safety Tool*. Oxford: Butterworth-Heinemann Ltd.

James, L.R., Demaree, R.G. and Wolf, G. (1993) 'r_{wg}: An assessment of within-group interrater agreement', *Journal of Applied Psychology*, 78, 2: 306–9.

Janes, C.L. (1979) 'Agreement measurement and the judgement process', *Journal of Nervous and Mental Diseases*, 167: 343–7.

Johnson, C.W. (2001) 'Models and the use of counter-factual reasoning in accident investigations', *Proceedings of Design, Specification and Verification of Interactive Systems, 13-15 June, 2001, GIST Technical Report G-2001-1*, Glasgow: University of Glasgow.

Johnson, W.G. (1980) *MORT Safety Assurance Systems*. New York: Marcel Dekker.

Johnson-Laird, P.N. (1983) *Mental Models: Towards a Cognitive Science of Language, Inference, and Consciousness*. Cambridge: Cambridge University Press.

Jones, E.E. and Davis, K.E (1965) 'From acts to dispositions: The attribution process in person perceptions', in L. Berkowitz (ed.) *Advances in Experimental Social Psychology Vol. 2*. New York: Academic Press.

Jones, E.E., Roch, L., Shaver, K.G., Goethals, G.R. and Ward, L.M. (1968) 'Pattern of performance and ability attribution: an unexpected primacy effect', *Journal of Personality and Social Psychology*, 10: 317–40.

Kelley, H.H. (1967) 'Attribution theory in social psychology' in D. Levine (ed.) *Nebraska Symposium on Motivation 15*. Lincoln: University of Nebraska Press.

Kelley, H.H. and Michela, J.L (1980) 'Attribution theory and research', *Annual Review of Psychology*, 31: 457–503.

Kidd, R.F. and Amabile, T.M. (1981) 'Causal explanations in social interaction: some dialogues on dialogue', in J.H. Harvey, W.J. Ickes and R.F. Kidd, (eds) *New Directions In Attribution Research, Vol. 3*. Hillsdale NJ: Erlbaum.

Kirwin, B.A. (1988) *A Comparative Study of Five Human Reliability Assessment Techniques in Human Factors and Decision Making: Their Influence on Safety and Reliability*. London: Elsevier Applied Science.

Kirwin, B. (1992a) 'Human error identification in human reliability assessment. Part 1: Overview of approaches', *Applied Ergonomics*, 23: 299–318.

Kirwin, B. (1992b) 'Human error identification in human reliability assessment. Part 2: Detailed comparison of techniques', *Applied Ergonomics*, 23: 371–81.

Kirwin, B. (1994) *A Practical Guide to Human Reliability Assessment*. London: Taylor and Francis.

Koch, S. (1964) 'Psychology and emerging conceptions of knowledge as unitary,' in T.W. Wann (ed.) *Behaviourism and Phenomenology*. Chicago: University of Chicago Press.

Komons, N.A. (1984) *The Cutting Air Crash: A Case Study in Early Federal Aviation Policy*. Washington: U.S. Department of Transportation.

Koran, L.M. (1975) 'The reliability of clinical methods, data and judgments', *New England Journal of Medicine*, 293: 642–6, 695–701.

Kosko, B. (1994) *Fuzzy Thinking*. London: HarperCollins.

Kress, G. (1990) 'Critical discourse analysis', in R. Kaplan (ed.) *Annual Review of Applied Linguistics*. New York: Cambridge University Press.

Kruglanski, A.E., Baldwin, M.W. and Towson, M.J. (1983) 'The lay epistemic process in attribution making', in M. Hewstone (ed.) *Attribution Theory: Social and Functional Extensions*. Oxford: Blackwell.

Kuhn, T.S. (1970) *The Structure of Scientific Revolutions*. Chicago: University of Chicago Press.

Lacroix, D. and Dejoy, D. (1989) 'Causal attribution to effort and supervisory response to workplace accidents', *Journal of Occupational Accidents*, 11: 97–109.

Lakoff, G. and Johnson, M. (1999) *Philosophy in the Flesh*. New York: Basic Books.

Lalljee, M. (1996) 'The interpreting self: an experimentalist perspective', in R. Stevens (ed.) *Understanding the Self*. London: Sage.

Lana, R.E. (1991) *Assumptions of Social Psychology: A Re-examination*. New Jersey: Erlbaum.

Larson, G.E. and Merrit, C.R. (1991) 'Can accidents be predicted? An empirical test of the cognitive failures questionnaire', *Applied Psychology*, 40, 1: 37–45.

Laurie, H. and Sullivan, O. (1991) 'Combining qualitative and quantitative data in the longitudinal study of household allocations', *The Sociological Review*, 39 (1): 113–30.

Lawson, H. and Apignanesi, L. (1989) *Dismantling Truth*. London: Weidenfeld and Nicolson.

Lecklund, L., Ingre, M., Kecklund, G., Söderström, M., Akerstedt, T., Lindberg, E., Jansson, A., Olsson, E., Sandblad, B. and Ålmqvist, P. (2001) 'The TRAIN Project: railway safety and the train driver information environment and work situation', Paper presented at Signalling Safety, London, 26–27 February. Online. Available   HTTP: http://www.hci.uu.se/projects/train/papers/signal ingsafety010214.pdf (accessed 11 September 2002).

Lee, M.D. and Del Fabbro, P.H. (2002) *A Bayesian Coefficient of Agreement for Binary Decisions*. Online. Available HTTP http://www.psychology. adelaide.edu.au/members/staff/michaellee/homepage/bayeskappa.pdf (accessed 20 September 2002).

Lehane, P. and Stubbs, D. (2001) 'The perceptions of managers and accident subjects in the service industries towards slip and trip accidents', *Applied Ergonomics*, 32: 119–26.

Leonard, T. and Hsu, J.S.J. (1999) *Bayesian Methods: An Analysis for Statisticians and Interdisciplinary Researchers*. New York: Cambridge University Press.

Leplat, J. and de Terssac, G. (1990) *Les Facteurs Humains de la Fiabilite Dans Les Systèmes Complexes*. Marseille: Octares.

Lewontin, R.C. (1993) *The Doctrine of DNA*. London: Penguin Books.

Lewycky, P. (1986) 'Toward an understanding of accident causes', *Canadian Occupational Safety*. September/October: 2–6.

Likert, R. (1932) 'A technique for the measurement of attitudes', *Archives of Psychology*, No. 140.

Locke, J. (1689) *A Letter Concerning Toleration*, trans. William Popple. Online. Available HTTP http://www.constitution.org/jl/tolerati.htm (accessed 20 September 2002).

Lofstedt, R. and Renn, O. (1998) 'The Brent Spar controversy: an example of risk communication gone wrong', in R. Lofstedt and L. Frewer (eds) *Risk and Modern Society*. London: Earthscan.

Logan, G.D. (1988) 'Toward an instance theory of automatization', *Psychological Review*, 95: 492–527.

Lozada-Larson, S.R. and Laughery, K.R. (1987) 'Do identical circumstances precede major and minor injuries?', in *Proceedings of the Human Factors Society 31st Annual Meeting*. Santa Monica, CA: The Human Factors Society.

Lucas, D.A. (1991) 'Organisational aspects of near miss reporting', in T.W. van der Schaaf, D.A. Lucas and A.R. Hale (eds) *Near Miss Reporting as a Safety Tool*. Oxford: Butterworth-Heinemann.

Lynn, R. (1966) *Attention, Arousal, and the Orientation Reaction*. Oxford: Pergamon Press.

McArthur, L.Z. and Post, D.L. (1977) 'Figural emphasis and person-perception', *Journal of Experimental Social Psychology*, 13: 520–35.

McBain, W.N. (1970) 'Arousal, monotony and accidents in line driving', *Journal of Applied Psychology*, 54, 6: 509–19.

McCauley, C. and Jacques, S. (1979) 'The popularity of conspiracy theories of presidential assassination: A Bayesian analysis', *Journal of Personality and Social Psychology*, 37: 637–44.

McClelland, J., Rumelhart, D. and the PDP Research Group (1986) *Parallel Distributed Processing: Explorations in the Microstructure Of Cognition, Volume 2: Psychological and Biological Models*. Cambridge, MA: MIT Press.

McClintock, C.G. (ed.) (1972) *Experimental Social Psychology*. New York: Holt, Rinehart and Winston.

McCullough, W. (1965) *Embodiments of Mind*. Cambridge, Mass: MIT Press.

Mackworth, J. (1969) *Vigilance and Habituation*. London: Penguin..

Mackworth, J. (1970) *Vigilance and Attention*. London: Penguin.

Maclure, M. and Willett, W.C. (1987) 'Misinterpretation and misuse of the Kappa statistic', *American Journal of Epidemiology*, 126, 2: 161–9.

Makin, P. and Sutherland, V. (1991) 'A fatal inversion?' *Occupational Safety and Health*, November 40–2.

Malcolm, N. (1977) *Memory and Mind*. London: Cornell University Press.

Mallery, J.C., Hurwitz, R. and Duffy, G. (1987) 'Hermeneutics: from textual explication to computer understanding?' in S. Shapiro (ed.) *The Encyclopaedia of Artificial Intelligence*. New York: John Wiley and Sons. Online. Available HTTP http://www.ai.mit.edu/people/jcma/papers/1986-ai-memo-871 (accessed 18 September 2002).

Manning, D.P. (1974) 'An accident model', *Occupational Safety and Health*, 4:1.

Margolis, H. (1987) *Patterns, Thinking and Cognition*. London: University of Chicago Press.

MARS (2002) (Marine Accident Reporting Scheme). Online. Available HTTP http://www.nautinst.org/marineac.htm (accessed 21 October 2002).

Martin, P. and Bateson, P. (1993) *Measuring Behaviour: An Introductory Guide*. Cambridge: Cambridge University Press.

Marx, D. (2001) 'Patient safety and the "just culture": a primer for health care executives', New York: Trustees of Columbia University. Online. Available HTTP http://www.mers-tm.net/support/Marx_Primer.pdf (accessed 21 October 2002).

Maslow, A.H. (1943) 'A theory of human motivation', *Psychological Review*, 50: 370–96.

Maxwell, A.E. (1977) 'Coefficients of agreement between observers and their interpretation', *British Journal of Psychiatry*, 130: 79–83.

Maxwell, J.A. (1998). 'Designing a qualitative study', in L. Bickman and D.J. Rog, (eds) *Handbook of Applied Social Research Methods*. London: Thousand Oaks.

Metzger, U. and Parasuraman, R. (2001) 'The role of the air traffic controller in future air traffic management: an empirical study of active control versus passive monitoring', *Human Factors*, 43, 4: 519–29.

Michotte, A. (1946) 'Á la perception de la causalité', in M.D. Vernon (ed.) *Experiments in Visual Perception*. London: Penguin.

Mikulincer, M. (1988) 'Reactance and helplessness following exposure to unsolvable problems: the effects of attributional style', *Journal of Personality and Social Psychology*, 54: 679–86.

Miles, M.B. (1979) 'Qualitative data as an attractive nuisance: the problem of analysis,' *Administrative Science Quarterly*, 24: 590–601.

Military Standard (1980) *Definitions of Terms for Reliability and Maintainability MIL STD 721 C*. New York: ANSI.

Miller, D.T. and Ross, M. (1975) 'Self-serving biases in the attribution of causality: Fact or fiction?', *Psychological Bulletin*, 82: 213–25.

Miller, G.A., Galanter, E. and Pribram, K.H. (1960) *Plans and the Structure of Behavior*. New York: Henry Holt.

Millward, L.J. (1995) 'Focus groups', in G.M. Breakwell, S. Hammond and C. Fife-Schaw (eds) *Research Methods in Psychology*. London: Sage.

Moles, A (1968) *Information Theory and Aesthetic Perception*. Urbana: University of Illinois Press.

Morris, C. (1938) *Foundations of the Theory of Signs*. Chicago: University of Chicago Press.

Munton, A.G., Silvester, J., Stratton, P. and Hanks, H. (1999) *Attributions in Action: A Practical Approach to Coding Qualitative Data*. Chichester: Wiley.

Murji, K. (1997) 'The agony and the ecstasy: drugs, media and morality', in R. Coomber (ed.) *The Control of Drugs and Drug Users*. Amsterdam: Harwood Academic Publishers.

National Health Service, Scotland (2002) *Learning from Experience*. Unpublished Document. Edinburgh: CRAG: St Andrews House.

Neisser, U. (1967) *Cognitive Psychology*. New York: Appleton-Century-Crofts.

Neisser, U. (1976) *Cognition and Reality*. San Fransisco: W.H. Freeman.

Newell, A. and Simon, H.A. (1972) *Human Problem Solving*. New Jersey: Prentice-Hall.

Newell A. and Simon, H.A. (1990) 'Computer science as empirical enquiry: symbols and search,' in Boden (ed.) *The Philosophy of Artificial Intelligence*. Oxford: Oxford University Press.

Niklasson, L. and van Gelder, T. (1994) 'Can connectionist models exhibit non-classical structure sensitivity?', in *Proceedings of the Sixteenth Annual Conference of the Cognitive Science Society*. Hillsdale, NJ: Lawrence Erlbaum Associates.

Nisbet, R.E. and Wilson, T.D. (1977) 'Telling more than we can know: verbal reports on mental processes', *Psychological Review*, 83, 3: 231–59.

Noë, Alva (2002) 'Is the visual world a grand illusion?' in *Journal of Consciousness Studies*, 9, 5–6: 1–12. Online. Available HTTP http://www.imprint.co.uk/pdf/ NOE.PDF (accessed September 2002).

Norman, D.A. (1983) 'Some observations on mental models', in D. Gentner and A. Stevens (eds) *Mental Models*. London: Lawrence Erlbaum.

Núñez, R. and Freeman W. (1999) *Reclaiming Cognition*. London: Imprint Academic.

Oaksford, M. (1993) 'Mental Models and the tractability of everyday reasoning, *Behavioral and Brain Sciences*, 16, 2: 360–1.

O'Brien, D.P., Braine, M.D.S. and Yang, Y. (1994) 'Propositional reasoning by mental models? Simple to refute in principle and in practice', *Psychological Review*, 101, 4: 711–24.

O'Hanlon, J.F. (1981) 'Boredom: practical consequences and a theory', *Acta Psychologica*, 49: 53–82.

O'Leary, M. (2001) 'The British Airways human factors reporting programme', *Reliability Engineering and System Safety*, 75: 245–55.

Palmer, R. (1969) *Hermeneutics*. Evanstong: Northwestern University Press.

Parker, I. (ed.) (1992) *Discourse Dynamics. Critical Analysis for Social and Individual Psychology*. London: Routledge.

Perrow, C. (1984) *Normal Accidents: Living With High-Risk Technologies*. Princeton, NJ: Princeton University Press.

Peschl, M. and Reigler, A. (1999) 'Does representation need reality', in A. Riegler, M. Peschl and A. von Stein (eds) *Understanding Representation in the Cognitive Sciences*, Dordrecht: Kluwer.

Peterson, C., Luborsky, L. and Seligman, M.E.P. (1983) 'Attributions and depressive mood shifts: a case study using the symptom-context method', *Journal of Abnormal Psychology*, 92: 96–103.

Peterson, C., Semmel, A., von Baeyer, C., Abramson, L.Y., Metalsky, G.I. and Seligman, M.E.P. (1982) 'The attributional style questionnaire', *Cognitive Therapy and Research*, 6, 3: 287–99.

Petersen, D.C. (1978) *Techniques of Safety Management*. New York: McGraw Hill.

Pidgeon, N.F. (1991) 'Safety culture and risk management in organisations', *Journal of Cross-Cultural Psychology*, 22: 129–40.

Popper, K. (1959) *The Logic of Scientific Discovery*. London: Hutchinson.

Posner, K.L., Sampson, P.D., Caplan, R.A., Ward, R.J. and Cheney, F.W (1990) 'Measuring inter-rater reliability among multiple raters: an example of methods for nominal data', *Statistics in Medicine*, 9: 1,103–15.

Potter, J. (1999) 'Post-cognitive psychology', in *Theory and Psychology*, 10, 1: 31–9. Online. Available HTTPhttp://www.lboro.ac.uk/departments/ss/depstaff/ staff/bio/JPpages/Post-cognitive%20psychology%20(web%20version).htm (accessed 20 September 2002).

Potter, J. and Wetherell, M. (1987) *Discourse and Social Psychology: Beyond Attitudes and Behaviour*. London: Sage.

Qin, Y. and Simon, H. (1995) 'Imagery and mental models in problem solving', in B. Chandrasekaran, J. Glasgow and N. Narayanan (eds) *Diagrammatic Reasoning*. London: AAAI Press.

Ramsay, W., Stich, S. and Garon. J. (1995) 'Connectionism, eliminativism, and the future of folk psychology', in C. Macdonald. and G. Macdonald (eds) *Connectionism: Debates on Psychological Explanation*. Oxford: Blackwell.

Rasmussen, J. (1976) 'Outlines of a hybrid model of the process plant operator,' in T.B. Sheridan and G. Johannsen (eds) *Monitoring Behavior and Supervisory Control*. London: Plenum Press.

Rasmussen, J (1978) *Notes on Diagnostic Strategies in Process Plant Environment Risø-M-1983*. Risø: Risø National Laboratory.

Rasmussen, J. (1979) *M 2192–On the Structure of Knowledge – A Morphology Of Mental Models in A Man-Machine Context*. Røskilde, Denmark: Risø National Laboratory.

Rasmussen, J. (1980a) 'The human as a systems component' in H.T. Smith and T.R.G. Green (eds) *Human Interaction with Computers*. London: Academic Press.

Rasmussen, J. (1980b) 'Models of mental strategies in process plant diagnosis' in J. Rasmussen (ed.) *Human Detection and Diagnosis of System Failures*. London: Plenum Press.

Rasmussen, J. (1980c) 'What can be learned from Human Error reports?', in K. Duncan, M. Gruneberg and D. Wallis (eds) *Changes in Working Life*. Chichester: John Wiley and Sons.

Rasmussen, J. (1982) 'Human errors. A taxonomy for describing human malfunction in industrial installations', *Journal of Occupational Accidents*, 4: 311–33.

Rasmussen, J. (1983) 'Skills, rules and knowledge; signals, signs, and symbols, and other distinctions in human performance models', *IEEE Transaction on Systems, Man, and Cybernetics*, 13, 3: 257–66.

Rasmussen, J. (1986) *Information Processing and Human-Machine Interaction*. New York: North-Holland.

Rasmussen, J. (1987a) 'The definition of human error and a taxonomy for technical systems design' in J. Rasmussen, K. Duncan and J. Laplat (eds) *New Technology and Human Error*. Chichester: John Wiley and Sons.

Rasmussen, J. (1987b) 'Cognitive control and human error mechanisms' in J. Rasmussen, K. Duncan and J. Laplat (eds) *New Technology and Human Error*. Chichester: John Wiley and Sons.

Rasmussen, J. (1990) 'Mental models and the control of action in complex environments', in D. Ackermann and M. Tauber (eds) *Human Factors in Information Technology*. Amsterdam: North-Holland.

Rasmussen, J., Duncan, K. and Leplat, J. (eds) (1987) *New Technology and Human Error*. Chichester, New York: John Wiley and Sons.

Rasmussen, J., Pedersen, O.M., Mancini, G., Carnino, A., Griffon, M. and Gagnolet, P. (1981) *Classification System for Reporting Events Involving Human Malfunctions*. Røskilde, Denmark: Risø National Laboratory.

Rauterberg, M. (1997) 'About faults, errors and other dangerous things', in C. Ntuen and E. Park (eds) *Human Interaction with Complex Systems: Conceptual Principles and Design Practice*. Norwell: Kluwer. Online. Available HTTP

http://www.ipo.tue.nl/homepages/mrauterb/publications/SHICS95paper.pdf (accessed 20 September 2002).

Reason, J. (1974) *Man in Motion: The Psychology of Travel.* London: Weidenfield and Nicolson.

Reason, J. (1990) *Human Error.* Cambridge: Cambridge University Press.

Reason, J. (1994) 'Forward' in M.S. Bogner (ed.) *Human Error in Medicine.* Hillsdale, NJ: Erlbaum.

Reason, J. (1997) *Managing the Risks of Organizational Accidents.* Aldershot: Ashgate.

Reason, J. (2001) *Human Error Reduction Operation.* (for Railway Safety Limited) England: The Vision Consultancy.

Renn, O., Webler,T. and Kastenholtz, H. (1998) 'Procedural and substantive fairness in landfill siting; a Swiss case study', in R. Lofstedt and L. Frewer (eds) *Risk and Modern Society.* London: Earthscan.

Rennie, D. (1999) 'Qualitative research: a matter of hermeneutics and the sociology of knowledge', in M. Koppala and L. Suzuki (eds) *Using Qualitative Methods in Psychology.* Thousand Oaks, CA: Sage.

Rescher, N., (1973) *The Coherence Theory of Truth.* Oxford: Oxford University Press.

Ricoeur, P. (1977) 'The model of the text: meaningful action considered as a text', in F. Dallmayr and T. McCarthy (eds) *Understanding and Social Inquiry.* London: University of Notre Dame Press.

Ricoeur, P. (1981) *Hermeneutics and the Human Sciences.* Cambridge: Cambridge University Press.

RIDDOR (1995) *Reporting Of Injuries, Diseases And Dangerous Occurrences Regulations.* United Kingdom Health and Safety Executive. Online. Available HTTP http://www.asu.org.uk/hands/riddor.html (accessed 15 October 2002).

Rimson, P.E. 'Investigating "causes"'(1998) Paper presented at the International Society of Air Safety Investigators International Seminar, Barcelona, Spain, 20 October.

Robson, C. (1993) *Real world research: A resource for social scientists and practitioner-researchers.* Oxford: Blackwell.

Roethlisberger, F.J. and Dickson, W.J. (1939) *Management and the Worker.* Cambridge, Mass: Harvard University Press.

Ross, A.J. (2003) 'Attributional style and performance', unpublished PhD thesis, University of Strathclyde, Glasgow.

Ross, A.J., Davies, J.B., White, M., Baxter, J., Wright, L. and Harris, J. (1999) 'Trend and pattern methodology for human factors root causes in events', *Proceedings of the British Psychological Society,* 7, 2: 116.

Ross, L. (1977) 'The intuitive psychologist and his shortcomings: distortions in the attribution process', in L. Berkowitz (ed.) *Advances in Experimental Social Psychology Vol. 10,* New York: Academic Press.

Rousseau, D.M. (1988) 'The construction of climate in organisational research', in C.L. Cooper, and I.T. Robertson (eds) *International Review of Industrial and Organisational Psychology.* Chichester: Wiley.

Royal College of Physicians (2000) *Nicotine Addiction in Britain.* London: RCP.

Rudisill, M. (1995) 'Line pilots' attitudes about and experience with flight deck automation: results of an international survey and proposed guidelines', in R.S. Jensen and L.A. Rakovan (eds) *Proceedings of the Eighth International Symposium on Aviation Psychology*. Columbus: The Ohio State University Press. Online. Available HTTP http://techreports.larc.nasa.gov/ltrs/PDF/NASA-ijap-95-mr.pdf (accessed 11 September 2002).

Rumelhart, D.E., Hinton, G.E., and McClelland, J.L. (1986) 'A general framework for parallel distributed processing', in D.E. Rumelhart, J.L. McClelland and the PDP Research Group (eds) *Parallel Distributed Processing: Explorations in the Microstructure of Cognition. Volume 1: Foundations*. Cambridge, MA: MIT Press.

Ryle, G. (2000) *The Concept of Mind*. London: Penguin.

Sacks, H. (1992) *Lectures on Conversation Vols. I and II*. Oxford: Blackwell.

Salkind, N. (2000) *Exploring Research*. New Jersey: Prentice-Hall.

Salomiemi, A. and Oksanen, H. (1998) 'Accidents and fatal accidents – some paradoxes', *Safety Science*, 29: 59–66.

Saludadez, J.A. and Garcia, P.G. (2001) 'Seeing our quantitative counterparts: construction of qualitative research in a roundtable discussion forum', *Qualitative Social Research* [on-line journal], 2, 1. Online. Available HTTP http://qualitative-research.net/fqs/fqs-eng.htm (accessed 20 September 2002).

Sanderson, P. and Harwood, K. (1988) 'The skills, rules and knowledge classification', in Goodstein, Anderson and Olsen (eds) *Tasks, Errors and Mental Models*. London: Taylor and Francis.

Sawyer, Alan G. (with Peter, J.) (1983) 'The significance of statistical significance testing in marketing research', *Journal of Marketing Research*, 20 (May): 122–33.

Schaler, J.A. (2000) *Addiction is a Choice*. Chicago: Open Court.

Schumann, H. and Presser, S. (1996) *Questions and Answers in Attitude Surveys: Experiments on Question Form, Wording and Context*. London: Sage.

Searle, J. (1990) 'Presidential address', *Proceedings of the American Philosophical Association*. Online. Available HTTP http://www.ecs.soton.ac.uk/~harnad/Papers/Py104/searle.comp.html (accessed 11 September 2002).

Seigler, R.S. and Liebert, R.M. (1974) 'Effects of contiguity, regularity and age on children's causal inferences', *Developmental Psychology*, 10: 574–9.

Seligman, M.E.P. (1979) 'The attributional style questionnaire', *Cognitive Therapy and Research* 6, 3: 287–300.

Shannon, C. (1948) 'A mathematical theory of communication', Bell System Technical Journal, 27, 379–423 and 623–56. Online. Available HTTP http://cm.bell-labs.com/cm/ms/what/shannonday/paper.html (accessed 11 September 2002).

Shannon, H.S. and Manning, D.P. (1980) 'Differences between lost-time and non-lost-time industrial accidents', *Journal of Occupational Accidents*, 2: 265–72.

Sharkey, A. (1996) 'Sorted or distorted?', *Guardian*, 26 Jan: 2–3.

Sherry, P. (2000) *Fatigue Countermeasures in the Railroad Industry: Past and Current Developments*. Washington: Association of American Railroads. Online. Available HTTP http://www.du.edu/transportation/fatigue/fatigue.pdf (accessed 11 September 2002).

Shorrock, S.T. and Kirwin, B. (In Press) 'Development and application of a human error identification tool for air traffic control', *Applied Ergonomics*.

Shrader-Frechette, K. (1998) 'Scientific method, anti-foundationalism and public decision making', in R. Lofstedt and L. Frewer (eds) *Risk and Modern Society*.London: Earthscan.

Shrivastava, P., Miller D. and Miglani, A. (1991) 'The evolution of crises: crisis precursors', *The International Journal of Mass Emergencies and Disasters*, 9, 3: 321–37.

Silvester, J., Anderson, N.R. and Patterson, F. (1999) 'Organizational Culture Change: An inter-group attributional analysis', *Journal of Occupational and Organisational Psychology*', 72: 1–23.

Sivia, D. S. (1996) *Data Analysis: A Bayesian Tutorial*. Oxford: Clarendon Press.

Skiba, R. (1985) *Taschenbuch Arbeitssicherheit*. Bielefeld: Erich Schmid Verlag.

Skinner, B.F. (1938) *The Behaviour of Organisms: An Experimental Analysis*. New York: Appleton-Century-Crofts.

Skinner, B.F. (1984) 'Representations and misrepresentations', *The Behavioral and Brain Sciences*, 7: 655–67.

Slovic, P. (1998) 'Perceived risk, trust and democracy', in R. Lofstedt and L. Frewer (eds) *Risk and Modern Society*. London: Earthscan.

Smith, R.P. (1981) 'Boredom: a review', *Human Factors*, 23(3): 329–40.

Spitznagel, E.L. and Helzer, J.E. (1985) 'A proposed solution to the base rate problem in the kappa statistic', *Archives of General Psychiatry*, 42: 725–8.

Staddon, J.R. (1999) 'Theoretical behaviorism', in W. O'Donohue and R. Kitchener (eds) *Handbook of Behaviorism*. London: Academic Press.

Stanton, N.A. and Stevenage, S.V. (1998) 'Learning to predict human error: issues of acceptability, reliability and validity', *Ergonomics*, 41, 11: 1,737–56.

Stern, D.G. (1991) 'Models of memory: Wittgenstein and cognitive science', *Philosophical Psychology*, 1991, 4, 2: 203–19. Online. Available HTTP http://www.geocities.com/wittgensteinonline/articles/memory.htm (accessed 4 April 2002).

Still, A. (1998) 'Theories of meaning', in R. Sapsford, A. Still, D. Miell, R. Stevens and M. Wetherell (eds) *Theory and Social Psychology*. London: Sage.

Stratton, P., Munton, A.G., Hanks, H.G.I., Heard, D.H. and Davidson, C. (1988) *Leeds Attributional Coding System (LACS) Manual*. Leeds: LFRTC.

Sutherland, V., Makin, P. and Cox, C. (2000) *The Management of Safety*. London: Sage.

Swain A.D. (1974) *The Human Element in Systems Safety: A Guide For Modern Management*. Camberley: InComtec Ltd.

Swain A.D. (1990) 'Human reliability analysis: need, status, trends and limitations' *Reliability Engineering and System Safety*, 29: 301–13.

Swain, A.D. and Guttman, H. E. (1983) *A Handbook of Human Reliability Analysis with Emphasis on Nuclear Power Plant Applications NUREG/CR-1278*. Washington: USNRC.

Tarski, A. (1944) 'The semantic conception of truth and the foundations of semantics', *Philosophy and Phenomenological Research*, 4. Online. Available HTTP http://www.ditext.com/tarski/tarski.html (accessed 19 September 2002).

Tarski, A. (1956) 'The concept of truth in formalized languages', in A. Tarski (ed.) *Logic, Semantics, Metamathematics*, Oxford: Clarendon Press.

Taylor, D.H. (1987) 'The hermeneutics of accidents and safety,' in J. Rasmussen, K. Duncan. and J. Leplat (eds) *New Technology and Human Error*. London: John Wiley and Sons Ltd.

Taylor, S.E. and Fiske, S.T. (1975) 'Point of view and perceptions of causality', *Journal of Personality and Social Psychology*, 32: 439–445.

Tejero, P. and Chóliz, M. (2002) 'Driving on the motorway: the effect of alternating speed on driver's activation level and mental effort', *Ergonomics*, 45, 9: 605–18.

Thagard, P., Eliasmith, C., Rusnock, P. and Shelley, C.P. (2002) 'Knowledge and Coherence', in R. Elio (ed.) *Common Sense, Reasoning, And Rationality (Vol. 11)*, New York: Oxford University Press. Online. Available HTTP http://cogsci.uwaterloo.ca/Articles/Pages/epistemic.html (accessed 19 September 2002).

Thompson, W.D. and Walter, S.D. (1988) 'A reappraisal of the k coefficient: k and the concept of independent errors', *Journal of Clinical Epidemiology*, 41: 949–58, 969–70.

Thorndike, R.L. (1951) 'Reliability', in D.N. Jackson, and S. Messick, (eds) *Problems in Human Assessment*. New York: McGraw-Hill.

Timpe, K.-P. (1993) 'Psychology's contributions to the improvement of safety and reliability in the man-machine system' in B. Wilpert and T. Qvale (eds) *Reliability and Safety in Hazardous Work Systems: Approaches to Analysis and Design*, Hove, U.K.: Lawrence Erlbaum Associates.

Tinline, G. and Wright, M.S. (1993) 'Further development of an audit technique for the evaluation and management of risk. Tasks 7 and 8', *Final report C2278: A Study for the Health and Safety Executive*, London: Four Elements.

Toft, B. (1996) 'Limits to the mathematical modelling of disasters', in C. Hood and D.K.C. Jones (eds) *Accident and Design: Contemporary Debates in Risk Management*. London: UCL Press.

Tomlinson, C. (1997) 'The PRICES approach to human error', in F. Redmill and T. Anderson (eds) *Safer Systems*. London: Springer-Verlag.

Turner B. and Pidgeon, N. (1998) *Man Made Disasters*. Oxford: Butterworth-Heinemann Ltd.

Turvey, M.T. and Carello, C. (1995) 'Some dynamical themes in perception and action,' in T.J. van Gelder, and R.F. Port (eds) *Mind as Motion*. London: MIT Press.

van der Schaaf, T.W. (1991) 'A framework for designing near miss management systems', in T.W. van der Schaaf, D.A. Lucas and A. R. Hale (eds) *Near Miss Reporting as a Safety Tool*. Oxford: Butterworth-Heinemann Ltd.

van Gelder, T.J. and Port, R.F. (1995) 'It's about time: An overview of the dynamical approach to cognition,' in T.J. van Gelder and R.F. Port (eds) *Mind as Motion*. London: MIT Press.

van Gelder, T.J. (1998) 'The dynamical hypothesis in cognitive science', *Behavioral and Brain Sciences*, 21: 1–14.

van Gelder, T.J. (1999a) 'Distributed versus local representation', in R. Wilson and F. Keil (eds) *The MIT Encyclopedia of Cognitive Sciences*. Cambridge, MA: MIT Press.

van Gelder, T.J. (1999b) 'Dynamic approaches to cognition' in R. Wilson and F. Keil (eds) *The MIT Encyclopedia of Cognitive Sciences*. Cambridge, MA: MIT Press.

Varela, F., Thompson, E. and Rosch, E. (1992) *The Embodied Mind: Cognitive Science and Human Experience*. London: MIT Press.

Vernon, M.D. (1962) *The Psychology of Perception*. London: University of London Press.

Vernon, M.D. (1970) *Perception Through Experience*. London: Methuen and Company.

Vicente, K., Christoffersen, K. and Pereklita, A. (1995) 'Supporting operator problem solving through ecological interface design', *IEEE Transactions on Systems, Man and Cybernetics*, 25, 4: 529–45.

Vicente, K. and Rasmussen, J. (1988) *A Theoretical Framework for Ecological Interface Design: Risø-M-2736*. Risø: Risø National Laboratory.

Vicente, K. and Rasmussen, J. (1992) 'Ecological interface design: theoretical foundations,' *IEEE Transactions on Systems, Man and Cybernetics*, 22, 4: 589–606.

Vitez, M., Karanyi, G., Gonczy, E., Rudas, T. and Czeizel, A. (1984) 'A semiquantitative score system for epidemiologic studies of fetal alcohol syndrome', *American Journal of Epidemiology*, 119: 301–8.

Wagenaar, W. and Groeneweg, J. (1987) 'Accidents at sea: multiple causes and impossible consequence', *International Journal of Man–Machine Studies*, 27, 5: 587–98.

Wagenaar, A., Groeneweg, J., Hudson, P.T.W. and Reason, J.T. (1994) 'Promoting safety in the oil industry', *Ergonomics* 37: 1,999–2,013.

Wagenaar, A., and van der Schrier, J. (1997) 'Accident analysis: the goal, and how to get there', *Safety Science*, 26, 1: 25–33.

Wallace, B., Ross, A., Davies, J.B., Wright, L. and White, M. (2002) 'The creation of a new minor event coding system', *Cognition Technology and Work*, 4: 1–8.

Ward, J., Mattick, R. and Hall, W. (1992) *Key Issues in Methadone Maintenance Treatment*. Kensington, NSW: New South Wales University Press.

Watson, J.B. (1919) *Psychology from the Standpoint of a Behaviorist*. Philadelphia: J.B. Lippincott Co.

Webb, G.R., Redmand, S., Wilkinson, C. and Sanson-Fisher, R.W. (1989) 'Filtering effects in reporting work injuries,' *Accident Analysis and Prevention*, 21: 115–23.

Weiner, B. (ed.) (1974a) *Achievement Motivation and Attribution Theory*. New Jersey: GL.

Weiner, B. (1974b) 'Achievement motivation as conceptualised by an attribution theorist', in B. Weiner (ed.) *Achievement Motivation and Attribution Theory*. New Jersey: GL.

Weiner, B. (1979) 'A theory of motivation for some classroom experiences', *Journal of Educational Psychology*, 71: 3–25.

Weiss, J. (1972) 'Psychological factors in stress and disease', *Scientific American*, 226, 6: 104–13.

Wetherell, M. (2001) 'Debates in discourse research', in M. Wetherell, S. Taylor and S.J. Yates (eds) *Discourse, Theory and Practice: A Reader*. London: Sage.

Wetherell, M. (ed.) (1996) *Identities, Groups and Social Issues*. London: Sage.

Wetherell, M. and Potter, P. (1988) 'Discourse analysis and the identification of interpretative repertoires', in C. Antaki (ed.) *Analysing Everyday Explanation: A Casebook of Methods*. London: Sage.

Wetherick, N.E. (1993) 'More models just means more difficulty', *Behavioral and Brain Sciences*, 16, 2: 367–8.

White, M.P. and Davies, J.B. (1998) 'The effects of context and sensitivity on self-reported attitudes towards drugs', *Journal of Substance Misuse*, 3: 213–20.

Whitehead, A.N. (1928) *Symbolism, Its Meaning and Effect*. Cambridge: Cambridge University Press.

Whorf, B.L. (1956) *Language, Thought and Reality*. Cambridge, MA: MIT Press.

Wickens, C. and Hollands, J.G. (2000) *Engineering Psychology and Human Performance*. New Jersey: Prentice-Hall.

Wiener, E.L. (1985) 'Beyond the sterile cockpit', *Human Factors*, 27, 1: 75–89.

Wiener, N. (1949) *Cybernetics*. New York: John Wiley.

Wilcox, S. and Katz, S. (1981) 'A direct realist alternative to the traditional conception of memory.', *Behaviorism*, 9: 227–39.

Wilde, G. (1994) *Target Risk*. Ontario: PDS Publications. Online. Available HTTP: http://pavlov.psyc.queensu.ca/target/ (accessed 17 October 2002).

Williams, J.C. (1986) HEART – A proposed method for assessing and reducing human error, in *Proceedings of the 9th Advances in Reliability Technology Symposium, University of Bradford, 4th April*, Culcheth: NCSR/UKAEA.

Wilpert, B. and Fahlbruch, B. (1996) 'Integrating human factors in event analysis in nuclear power plants (NPP)', *International Journal of Psychology*, 31: 3–4, 2,394.

Wilpert, B. and Qvale, T. (1993) (eds) *Reliability and Safety in Hazardous Work Systems: Approaches to Analysis and Design*. Hove, U.K.: Lawrence Erlbaum Associates.

Winograd, T. and Flores, F. (1986) *Understanding Computers and Cognition: A New Foundation for Design*. Norwood: Ablex Publishing.

Wittgenstein, L. (1953) *Philosophical Investigations*. Oxford: Blackwell.

Wittgenstein, L. (1958) *The Blue and Brown Books*. Oxford: Blackwell.

Woods, D.D., Johannesen, L.J., Cook, R.I. and Sarter, N.B. (1994) *Behind Human Error: Cognitive Systems, Computers and Hindsight*. Columbus, Ohio: CSE-RIAC.

Wright, L.B. (2002) 'The analysis of UK railway accidents and incidents: a comparison of their causal patterns'. Unpublished PhD thesis, University of Strathclyde, Glasgow.

Wright, L.B., Davies J.B., Courtney E. and Reid H. (2000) 'CIRAS: collecting and analysing human factors data from the UK rail industry', in M. Cottam, D. Harvey, R. Pape, R. Tait and J. Tait (eds) *Proceedings of ESREL 2000, SARS and*

SRA-Europe Annual Conference, Foresight and Precaution. Rotterdam: Balkema.

Yerkes, R.M. and Dodson, J.D. (1908) 'The relation of strength of stimulus to rapidity of habit-formation', *Journal of Comparative Neurophysiological Psychology*, 18: 459–82.

Yule, G.U. (1912) 'On the methods of measuring association between two attributes', *Journal of the Royal Statistical Society*, 75: 581–642.

Zajonc, R.B., Heingartner, E.M. and Herman, E.M. (1969) 'Social enhancement and impairment of performance in the cockroach', *Journal of Personality and Social Psychology*, 13: 83–92.

Zuckerman, M. (1979) *Sensation Seeking.* Hillsdale: Lawrence Erlbaum.

Zuckerman, M. (1994) *Behavioral Expressions and Biosocial Bases of Sensation Seeking.* Cambridge: Cambridge University Press.

Index

For Product Safety Concerns and Information please contact our EU
representative GPSR@taylorandfrancis.com
Taylor & Francis Verlag GmbH, Kaufingerstraße 24, 80331 München, Germany

www.ingramcontent.com/pod-product-compliance
Ingram Content Group UK Ltd.
Pitfield, Milton Keynes, MK11 3LW, UK
UKHW021607240425
457818UK00018B/435